Data Response Questions in Advanced Level Geography

Physical Geography

Paul Guinness M.Sc.

Head of Geography, The London Oratory School

and

Kevin Ball M.Sc.

Head of Geography, Langdon Comprehensive School

HODDER AND STOUGHTON

LONDON SYDNEY AUCKLAND TORONTO

Contents

Biogeography

Introduction

In recent years data response questions have become a part of A-level examinations set by an increasing number of examination boards and have usually marked significant changes in syllabus content. The emergence of this style of question at A-level is also part of a general move to improve the assessment of a candidate's grasp of the basic concepts upon which modern geography is based. The role of factual recall which tends to dominate the traditional essay answer is deliberately kept to a minimum. Equally, data response questions are designed to test the ability to understand, analyse and apply data and basic techniques in a variety of different forms.

As the style of question differs markedly from the traditional long essay, so the technique required to respond successfully to the format of such structured questions is also different. Practice at tackling data response exercises is thus a vital part of the preparation required for the A-level examination. The aim of this book and its companion, *Data Response Questions in Advanced Level Geography: Human Geography*, is to provide the candidate with a variety of exercises testing a range of concepts, skills and techniques.

Ideally data response questions should be used over the full duration of the A-level course so that confidence in handling such exercises is gradually acquired. Although the book can also be used as a revision source in the months prior to the examination, a longer familiarity with this style of question is advised.

The data response questions set by the examination boards are of two general types: (i) limited data with restricted answer space (e.g. London Board) and (ii) more extensive data where no limitations are placed on the length of the answer (e.g. Joint Matriculation Board). Both varieties of question are contained in this text.

The book falls into four parts covering the elements of geomorphology, hydrology, meteorology and climatology, and biogeography. Each part is further subdivided into Sections A and B. The questions in Section A are of the limited data – restricted answer type while those in Section B contain more extensive data and require more detailed answers.

Following the format set by the examination boards the exercises in Section A are designed for completion in 15–20 minutes and those in Section B in 35–45 minutes.

For the restricted response questions a guide has been given as to the allocation of marks and the maximum number of lines for the answer for each part of the question, e.g.,

[5L–8M] = 5 lines, 8 marks.

Where a part of a question is further subdivided, (2 × [4L–4M]) = 4 lines, 4 marks for each of the two subdivisions.

In Section B, only a suggested mark allocation is provided to retain parity with current examination board policy (i.e., the Joint Matriculation Board). In line with the examination boards the maximum mark for all questions has been set at 25.

The space and mark allocations, finalised after careful testing with A-level students, are offered only as a guide and teachers may wish to amend them where a change in question emphasis is desired.

Preference has been given to British examples as these tend to form the basis of the teaching syllabus in physical geography.

Frequent reference is made to the use of fieldwork techniques, particularly with regard to the practical problems involved.

Geomorphology – Section A

I

(a) Explain why the two types of plate boundary shown can be termed constructive and destructive. $(2 \times [4L-4M])$

(b) Name and locate an area where each of these types of plate boundary can be found. $(2 \times [1L-2M])$

(c) Name the type of volcano associated with each of these plate boundaries. $(2 \times [1L-2M])$

(d) Draw an annotated diagram to show the essential characteristics of *one* of the volcanoes named in (c). $(9M)$

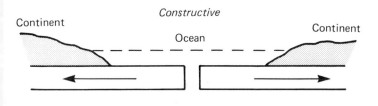

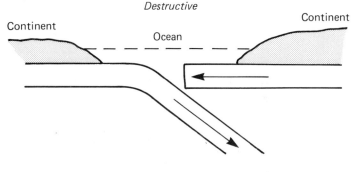

Plate boundaries and volcanic landforms

2

(a) What do you understand by the terms Marine Transgression and Alpine Orogeny? $(2 \times [2L-2M])$

Briefly explain their effects on the Weald. $(2 \times [3L-3M])$

(b) Describe the type of accordant drainage that you would expect to develop in this area after the Alpine Orogeny. $(4L-5M)$

(c) Explain how the original surface of the Weald has been modified into the scarp-ridge-vale topography of today. $(8L-10M)$

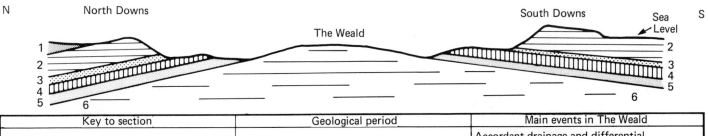

Key to section		Geological period	Main events in The Weald
		Present day	Accordant drainage and differential denudation have formed a scarp-vale landscape
		Oligocene-Miocene	Alpine Orogeny
1	Sand + clays	Eocene	
2	Chalk	Upper Cretaceous	Periods of uplift and marine transgression
3	Gault clay		
4	Lower greensand	Lower Cretaceous	
5	Weald clay		
6	Hastings beds		

Diagrammatic cross-section across the Wealden area of south-eastern England

3

(a) Comment on the sequence of geological events associated with the Caledonian and Hercynian Orogenies. *(6L–8M)*

(b) What evidence is there in southern England of the Alpine Orogeny? *(5L–5M)*

(c) Describe how the surface geology of Dartmoor reveals evidence of the volcanic intrusion 280–300 million years ago. *(6L–7M)*

(d) What geological evidence is there to suggest that the latitude of the British Isles was once much further south? *(5L–5M)*

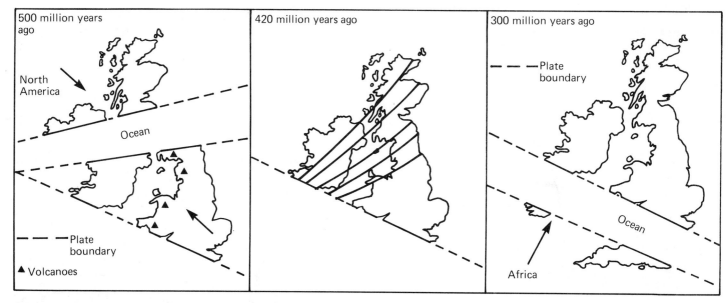

Figure 1   Ordovician Igneous and Metamorphic activity. Latitude 30° S

Figure 2   Caledonian Orogeny. Latitude 25° S

Figure 3   Upper Carboniferous. Latitude 0°

Figure 4   Hercynian Orogeny. Intrusive Volcanic and Metamorphic activity in south-west. Latitude 5° N

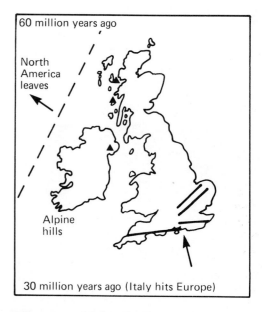

Figure 5   Alpine Orogeny. Latitude: present position 50–60° N

4

The photograph shows an area of the Scottish Highlands. The main valley is 90 km in length and traverses the area in a NE–SW direction. The lake shown, Lake Lochy, is 16 km long and is over 150 metres deep in places. The flat ground in the fore-ground is less than 35 metres above sea level and is largely composed of intrazonal soils. A canal has been built to the north of the River Lochy.

(a) Choose *two* landforms shown on the photograph, *one* resulting from tectonic activity, *the other* from glaciation. For each landform, briefly explain its formation, indicating where on the photograph it is found. (2 × [6L–6M])

(b) Describe the valley of the River Lochy shown in the foreground. (5L–6M)

(c) Briefly comment on the interrelationship between physical and human geography shown in the photograph. (6L–7M)

5

(a) In which stage of the cycle would you expect to find (i) gorges and scars, (ii) clints and grikes, (iii) sink holes and solution hollows? (*3L–3M*)

(b) Describe the conditions which are necessary for the Karstic cycle to operate on a landscape. (*8L–10M*)

(c) Briefly describe the chemical weathering processes which operate in areas of carboniferous limestone. (*6L–6M*)

(d) Explain why Cvijic's cycle, developed in the Yugoslavian karst region, has only limited application to British examples. (*4L–6M*)

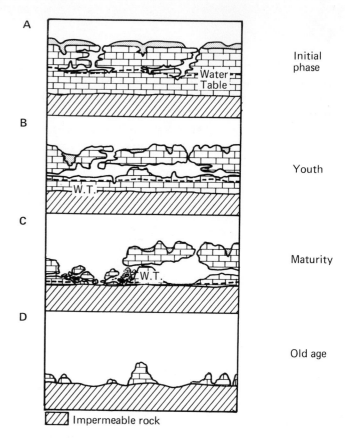

A — Initial phase — Water Table

B — Youth — W.T.

C — Maturity — W.T.

D — Old age

⬜ Impermeable rock

The Karstic cycle of erosion

6

(a) Explain the use of the term 'escarpment' in this area of the North Downs. (*3L–3M*)

(b) (i) Distinguish between a combe and a dry valley. (*4L–4M*)

    (ii) Suggest *two* possible modes of formation of dry valleys. (*2 × [4L–4M]*)

(c) Account for the location of the zone of periglacial landforms near the foot of the scarp slope. (*4L–6M*)

(d) Suggest reasons for the linear distribution of villages shown on the sketch-map. (*4L–4M*)

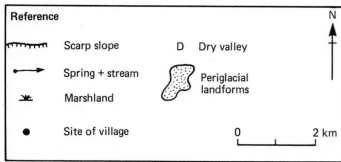

Reference

N

〰️ Scarp slope   D Dry valley

•→ Spring + stream   🟤 Periglacial landforms

⚘ Marshland

• Site of village   0 — 2 km

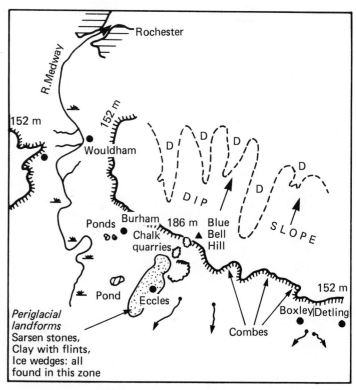

Rochester

R. Medway

152 m   152 m

Wouldham

D D D D D D

DIP   SLOPE

Ponds   Burham   186 m   Blue Bell Hill

Chalk quarries

Pond   Eccles

Periglacial landforms Sarsen stones, Clay with flints, Ice wedges: all found in this zone

Combes

152 m

Boxley Detling

Geomorphological features of the North Downs escarpment

7

(a) Match A, B, C and D on Figure 1 with the following sections normally found on a slope profile: (i) free-face, (ii) waxing slope, (iii) waning slope, (iv) rectilinear section. (4L–4M)

(b) Account for the upper convexity and the lower concavity of the slope profile. (8L–9M)

(c) Explain the *two* methods of slope development shown in Figure 2. (2 × [6L–6M])

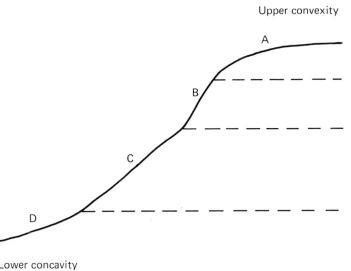

Figure 1   The elements of a slope profile

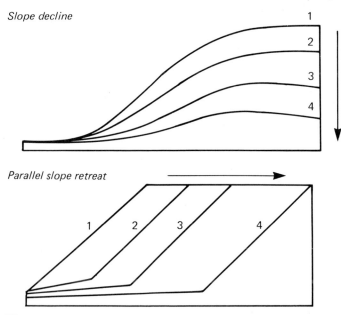

Figure 2

8

(a) Explain the pattern of weathering regions shown in the two graphs. (2 × [6L–5M])

(b) Describe the other variables which need to be considered when studying rates of weathering. (8L–10M)

(c) Explain why frost action rather than exfoliation is regarded as a more accurate indicator of mechanical weathering processes. (5L–5M)

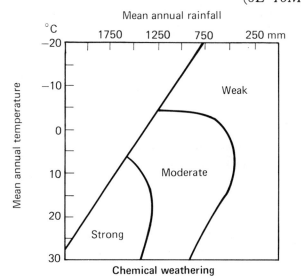

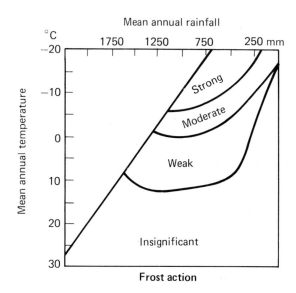

5

9

(a) Briefly distinguish between Flow, Slide and Fall as types of mass movement. (6L–6M)

(b) Slide movements usually involve the presence of a 'shear surface'; explain how this may be brought about. (4L–6M)

(c) Which factors are likely to affect the volume and speed of creep on a slope? (6L–8M)

(d) Explain how man can initiate mass movement on slopes. (6L–5M)

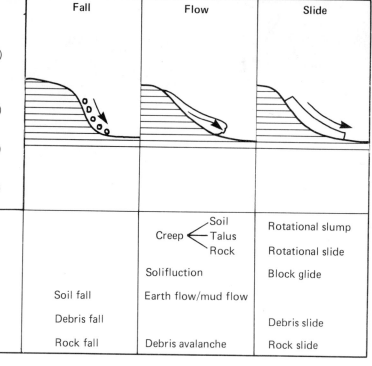

A classification of mass movements

10

(a) Name the type of mass movement shown in the section. (1L–2M)

(b) Describe the likely sequence of events leading up to the landslip of 1915. (8L–12M)

(c) Suggest a reason for the 1940 landslip, given that a different external factor was involved. (2L–3M)

(d) Explain how man can check or prevent mass movement such as this in areas similar to Folkestone Warren. (6L–8M)

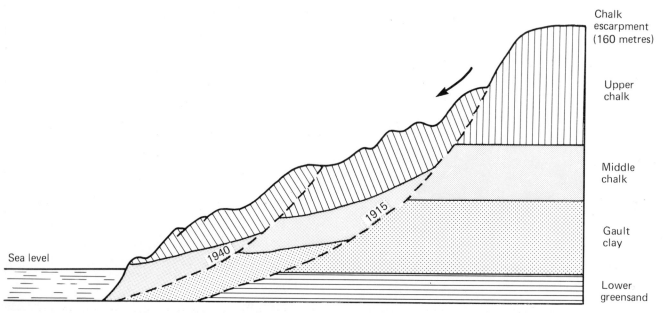

Diagrammatic section of the Folkestone Warren landslips of 1915 and 1940

II

(a) Account for the variety of beach material found on Slapton Sands. (6L–6M)

(b) Identify and explain *three* possible elements in the construction of Slapton Sands, besides the work of the sea. (3 × [5L–3M])

(c) Describe the type of vegetation that you would expect to find in Slapton Ley. (5L–6M)

(d) Explain the significance of the 130 metre summit accordance to geomorphologists. (3L–4M)

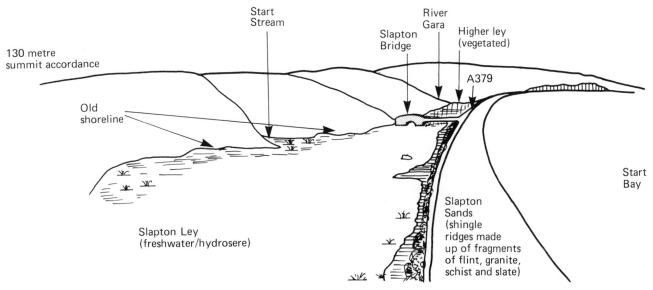

Figure 1   A student's field sketch of Slapton Sands

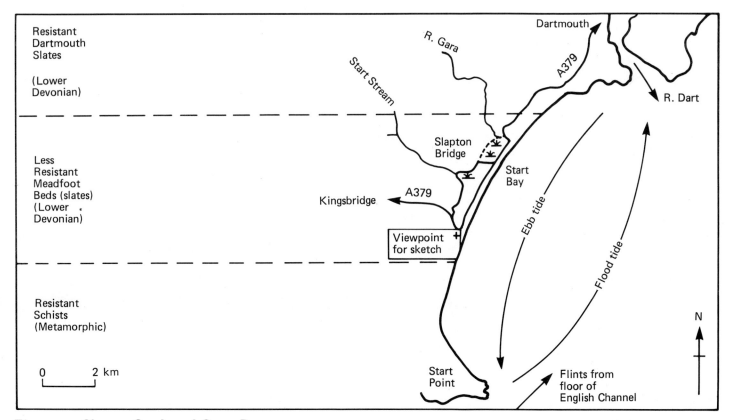

Figure 2   Slapton Sands and Start Bay

12

(a) Describe the geographical conditions necessary to give the type of coastal situation shown in the initial stage. (4L–4M)

(b) Describe the processes and landforms that are likely to occur in the intermediate stage. (10L–9M)

(c) Draw annotated diagrams, *both in plan and in profile*, to show the coastal forms which are likely to occur in the final stage of development. (2 × 6M)

The marine cycle of a submerged upland coast

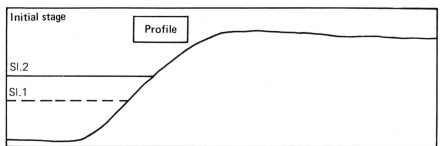

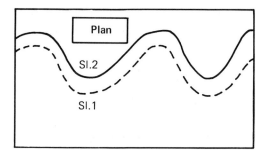

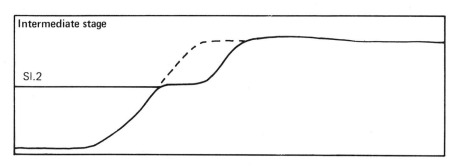

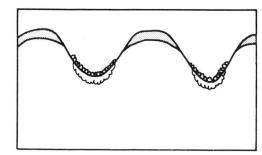

13

(a) Name the coastal feature that has evolved at Dungeness. (1L–2M)

(b) Describe the sequence of events which has led to the construction of this landform. (8L–11M)

(c) Explain the significance of: (i) the position of the 10 fathom line, (ii) the variation and direction of wave fetch. (8L–8M)

(d) What effect do the various depositional features on the coast appear to have had on the drainage pattern? (2L–4M)

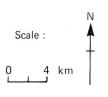

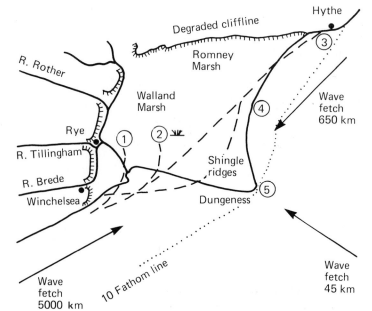

The evolution of Dungeness, Kent

14

(a) Explain the contribution made by the Pleistocene glaciation to the coastline of highland Britain.

(6L–8M)

(b) Describe the results of the post-glacial isostatic adjustment on: (i) the west coast of Scotland, (ii) the Thames Estuary. (2 × [3L–5M])

(c) Account for the relatively high amount of money spent on coastal defence work between Spurn Head and Dungeness. (6L–7M)

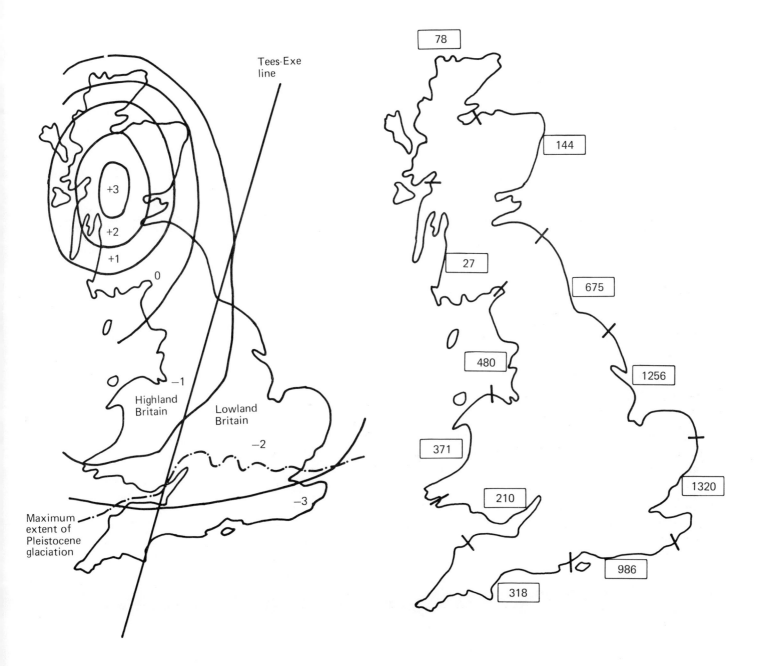

Post-glacial isostatic adjustment in Britain (relative movement of land in mm per year)

Cost of coastal defence work in Britain (£000 per year)

15

(a) Explain the pattern of cirque distribution shown in the table in Figure 1. (8L–9M)

(b) Abrasion, freeze-thaw, joint-block removal and rotational slip are four processes closely linked with the formation of the characteristic cirque profile. On a copy of Figure 2, indicate where each of these processes is important. (4M)

Aspect of cirques in the West-Central Lake District	
Aspect	Per cent facing each direction
NE	52.0
SE	19.2
NW	23.3
SW	5.5

Figure 1

(c) Explain the contribution of each of these four processes in the formation of the cirque basin. (8L–12M)

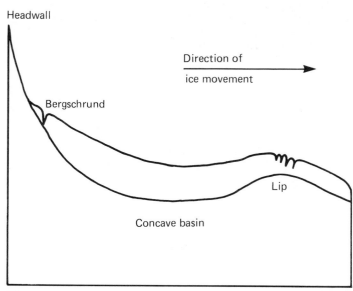

Figure 2   A cirque glacier

16

(a) Explain the terms Congelifraction and Solifluction and describe their function in the youthful stage of the periglacial cycle. (2 × [4L–3M])

(b) Describe the climatic conditions which characterise periglacial areas. (6L–5M)

(c) Name the landforms A, B and C shown in the mature stage of the cycle. Outline the common features of their development. (8L–10M)

The cycle of periglacial erosion

Youth

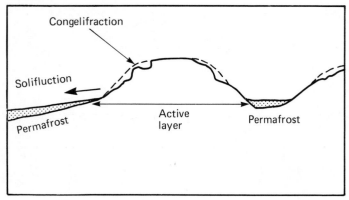

(d) Define the term Cryoplanation and locate one area where a cryoplanation surface might be found. (3L–4M)

Maturity

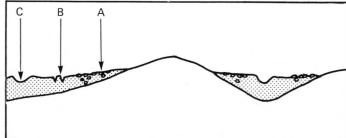

Old age

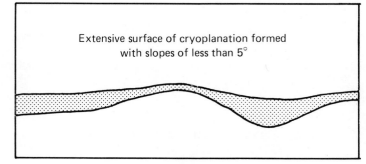

17

(a) Describe the nature and effects of the following processes in the evolution of arid landscapes:
Jey 31. (i) exfoliation, (ii) flash flooding, (iii) fluvial *P52* sorting, (iv) slope retreat, (v) aeolian processes. *P54 P55* Indicate, on a copy of the section shown, where these processes will operate most effectively.

(5 × [3L−3M])

(b) Assess the effect on the landscape of a period of prolonged aridity. *P72 Hutton P54. Wey* (5L−5M)

(c) Explain briefly why parallel slope retreat is likely to occur in tropical arid areas. (4L−5M)

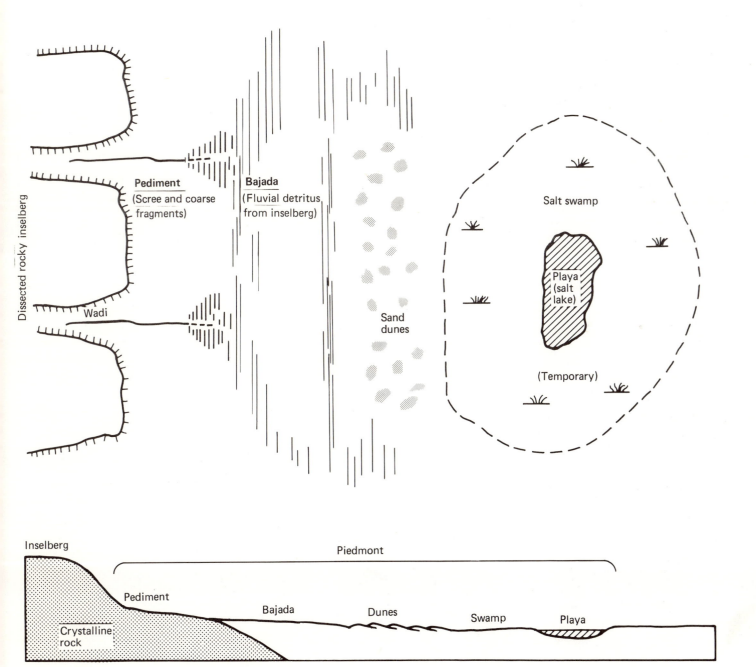

Descriptive plan of an arid landscape

18

(a) The hydraulic radius is a measure of stream efficiency.
   (i) Explain how it is calculated. (*1L–2M*)

   (ii) Describe and account for the pattern shown in the table. (*5L–7M*)

(b) Explain the pattern of meander amplitudes shown in the table, suggesting a reason for the lower values at survey points 5, 6 and 7. (*6L–6M*)

(c) Account for the variation in channel cross-profiles found near survey points 1 and 8. (*6L–6M*)

(d) Discharge is an integral part of stream hydraulics. Describe *two* ways in which it can be measured in the field. (*4L–4M*)

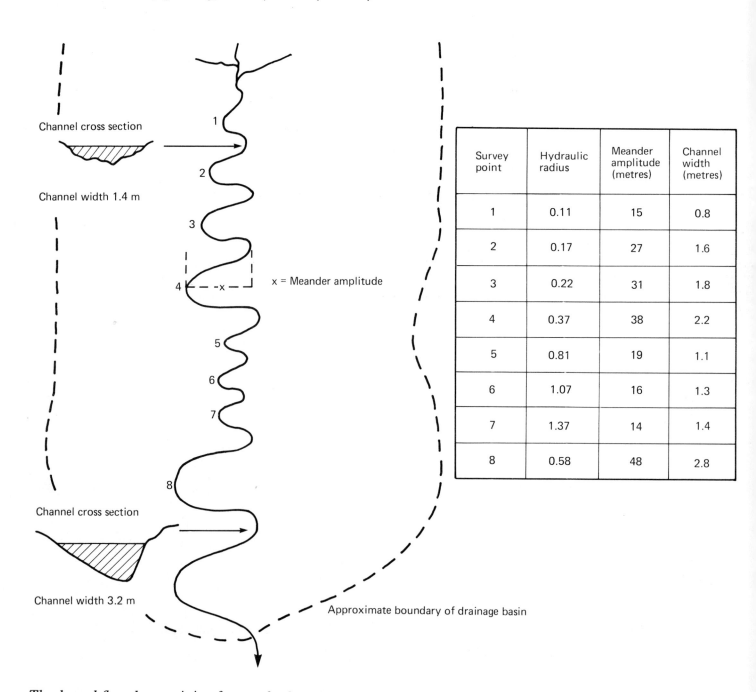

Survey point	Hydraulic radius	Meander amplitude (metres)	Channel width (metres)
1	0.11	15	0.8
2	0.17	27	1.6
3	0.22	31	1.8
4	0.37	38	2.2
5	0.81	19	1.1
6	1.07	16	1.3
7	1.37	14	1.4
8	0.58	48	2.8

Channel cross section

Channel width 1.4 m

x = Meander amplitude

Channel cross section

Channel width 3.2 m

Approximate boundary of drainage basin

The channel-flow characteristics of a second order stream in the Lake District

19

(a) Describe the features shown on the map at 1, 2 and 3 and explain how they support the hypothesis that river capture has taken place.

$(3 \times [4L-3M])$

(b) Draw a sketch-map to suggest the drainage pattern before river capture took place. $(5M)$

(c) Describe the likely characteristics of rock type B, suggesting an example. $(2L-2M)$

(d) Describe any further pieces of evidence which may indicate that the drainage has been diverted.

$(6L-9M)$

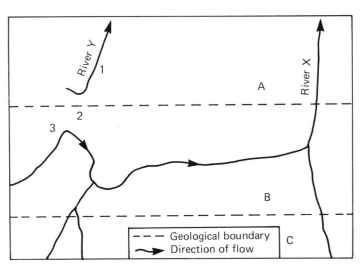

River capture

20

(a) The lower course of the River Wye appears to bear little relationship to geological structure.
(i) What name is given to this type of drainage? $(1L-2M)$

(ii) Briefly explain how it could have evolved. $(5L-4M)$

(b) Using Figure 1 as a guide, explain the occurence of the abandoned meanders at Newlands and St Briavels. $(6L-6M)$

(c) What is the name given to the type of meander shown in Figure 2? Describe how it could have been formed. $(6L-8M)$

(d) Explain why this section of the Wye Valley is untypical of a river in its lower course. $(6L-5M)$

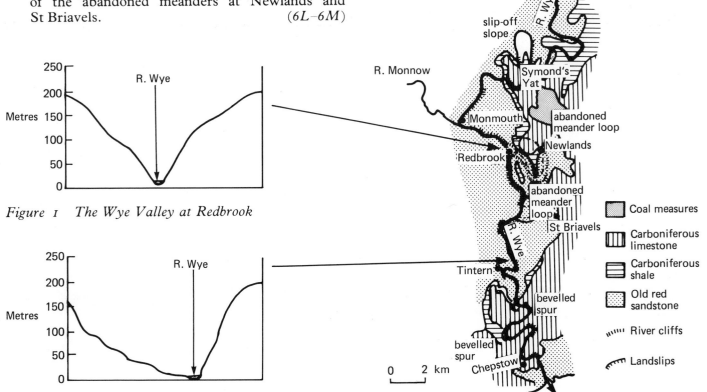

Figure 1   The Wye Valley at Redbrook

Figure 2   The meander loop at Tintern

The lower course of the River Wye

Geomorphology – Section B

21

'On 18 May 1980 a magnitude 5.0 earthquake triggered off the eruption of Mt St Helens. Around 2 cubic kilometres of material on the north slope was turned into a vast landslide. Everywhere within a 20 mile radius was devastated. Sixty-one people were either dead or missing. Some 2 million birds, fish and animals perished, 26 lakes were destroyed and 3.2 billion board feet of timber was blown down or destroyed. Overall farming losses were projected at 175 million dollars.' (*National Geographic Magazine*, January 1981)

(a) Name plates A and B and state the type of plate boundary that they help to create. (*3M*)

(b) Explain the role of the Juan De Fuca plate at this plate boundary. (*2M*)

(c) Explain, with the aid of diagrams, a possible mode of formation for Crater Lake (a caldera). (*7M*)

(d) Explain why this area of the world has come to be regarded as a laboratory for studying plate tectonics. (*8M*)

(e) Briefly outline any beneficial aspects of plate tectonics to human activity. (*5M*)

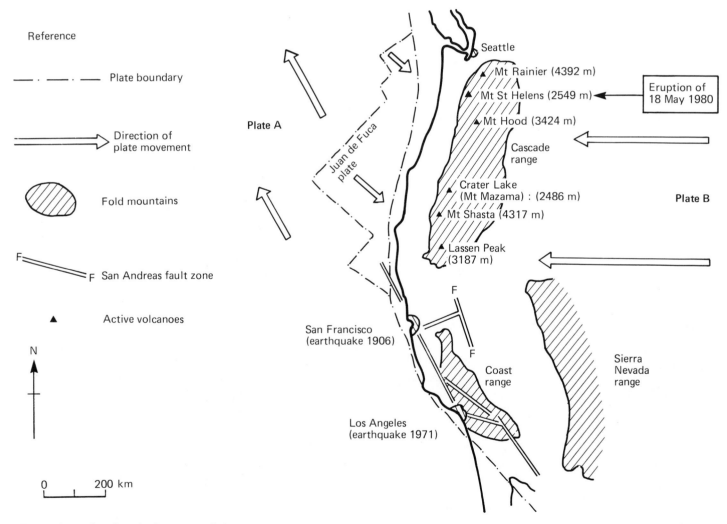

Tectonic and volcanic features of the western coast of the USA

14

22

(a) What is the name given to this type of coastline where the structural grain is at a high angle to the sea? *(2M)*

(b) Name and explain the process which is shown by arrows to be attacking the headlands. *(5M)*

(c) Account for the variety of erosional and depositional landforms shown on the sketch-map. *(10M)*

(d) What explanation, other than differential erosion, could you put forward to explain the nature of this stretch of coastline? *(3M)*

(e) There is a slight movement of material on Woolacombe Sands to the south. Briefly explain how you would set up a fieldwork experiment to examine the rate and direction of this longshore drift. *(5M)*

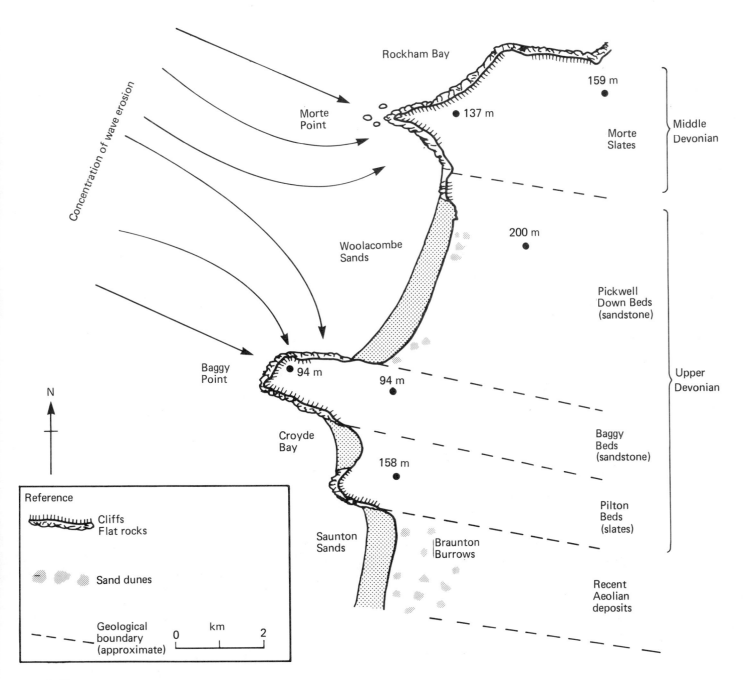

North Devon

23

(a) Complete the stream ordering of the two drainage basins shown according to the Strahler system. (It is suggested that students make a tracing of the original.) (4M)

(b) Calculate the number of streams in each stream order and then work out: (i) the stream intensity, (ii) the bifurcation ratios shown in the table. (9M)

(c) Explain how the drainage density of a drainage basin would be calculated. (2M)

(d) *Stream A*: Located in a non-glaciated granite upland with a maximum elevation of 621 metres.

Stream B: Located in a glaciated upland area with a maximum elevation of 977 metres. Parent rock – Borrowdale Volcanics.

Using the above information as a guide, describe and account for the main differences between the two drainage basins. (10M)

Stream A – Dartmoor

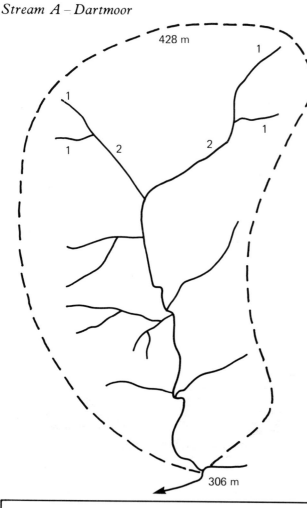

Stream B – The Lake District

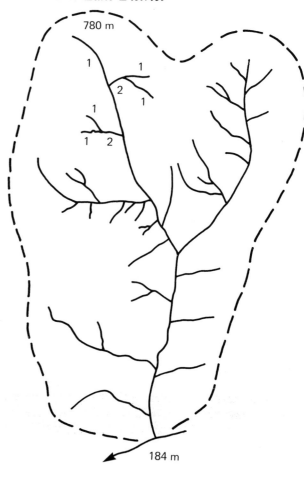

Drainage basin : 4th order Area of basin : 12.9 sq km Basin relief : 122 metres Total Number of Streams : 21 s.o.1 : .... s.o.3 : .... s.o.2 : .... s.o.4 : .... Stream intensity : .... Bifurcation Ratios : 1:2 = 　　　　　　　　　2:3 = 　　　　　　　　　3:4 =	Drainage basin : 4th order Area of basin : 12.7 sq km Basin relief : 596 metres Total Number of Streams : s.o.1 : .... s.o.3 : .... s.o.2 : .... s.o.4 : .... Stream intensity : .... Bifurcation Ratios : 1:2 = 　　　　　　　　　2:3 = 　　　　　　　　　3:4 =

24

(a) State in which zone of the model the following processes and landforms are most likely to be found.

Processes	Landforms
Solifluction	Drumlins
Nivation	Kame Terraces
Rotational Slip	Ice Wedges
Stratification of Drift	Roche Moutonnees
Ablation	Ribbon Lakes (5M)

(b) Explain the relationship that exists between the glaciated area (A,B,C,) and each of the peripheral zones D,E,F. (3 × 2M)

(c) Briefly account for the variety of landforms to be found in zone C, stating an area of the British Isles where these landforms could be studied in the field. (9M)

(d) Briefly comment on the limitations of the model and the attempt to represent spatial zones of glacial activity. (5M)

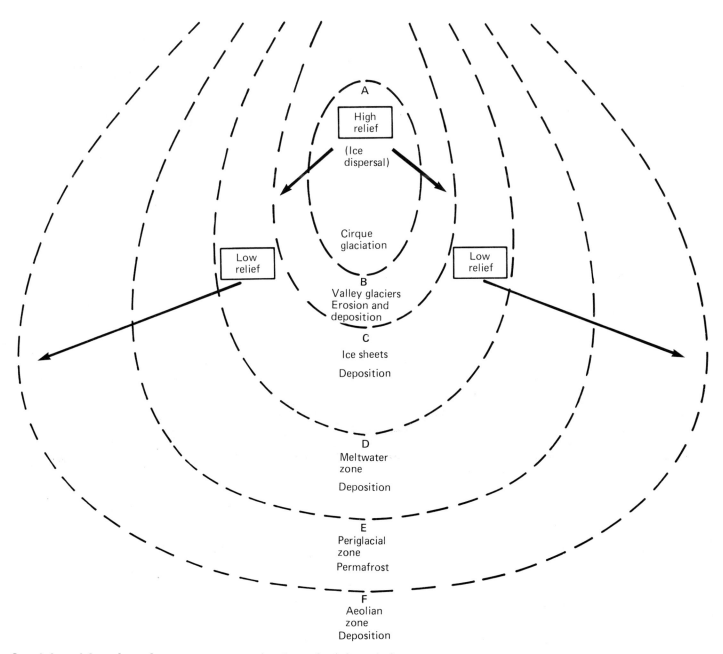

Spatial model to show the processes operating in a glacial period

25

The Rocky Tors of Dartmoor are the most well known of all granite landforms yet their formation is still a matter of some controversy. Their evolution has been attributed to deep chemical weathering, periglacial activity and the results of prolonged pediplanation.

(a) Describe how the characteristics of the granite itself appears to play a vital role in the formation of Haytor. (6M)

(b) The enlarged joints shown in the photograph contain deposits of growan (decomposed granite). Describe the weathering processes responsible. (8M)

(c) Explain why the evidence at Haytor suggests that the periglacial period was of vital importance in the formation of tors. (5M)

(d) Suggest a reason for the irregular pattern of jointing found on Haytor (4M)

(e) Tors also occur in Millstone Grit and Dolerite. Which quality does this suggest must be present in the parent rock from which these landforms are created? (2M)

Haytor, viewed from Haytor Rocks

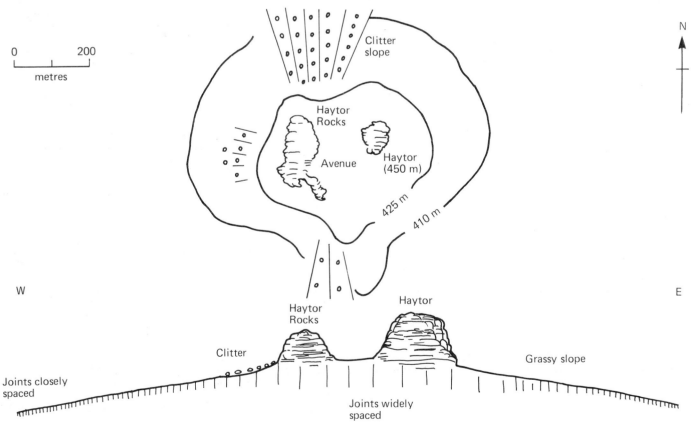

A sketch map and cross-section to show the granite features of the area

Hydrology – Section A

26

(a) Explain the terms (i) evapotranspiration, (ii) soil moisture, (iii) groundwater. $(3 \times [3L-2M])$

(b) Account for the fact that the land surface of the earth has a positive balance in the relationship between precipitation and evapotranspiration compared to the oceans. $(10L-10M)$

(c) What are the main factors that cause spatial variations in evapotranspiration rates over the land area of the earth? $(10L-9M)$

The hydrological cycle and water storage of the globe. The exchanges in the cycle are referred to 100 units which equal the mean annual global precipitation of 85.7 cm.

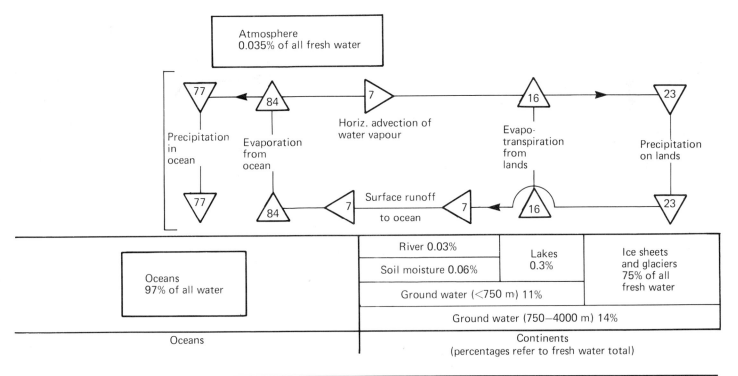

27

(a) What general differences might be expected in the contribution of precipitation to streamflow in each of the four zones? $(12L-10M)$

(b) Account for the changing level of soil saturation. $(7L-7M)$

(c) What effects would a change in land use to arable farming have on overland flow? $(7L-8M)$

A forested slope with uniform soil

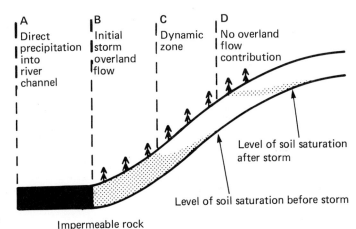

28

(a) How might such measurements of streamflow be calculated? *(4L–4M)*

(b) Attempt to explain the pattern of precipitation over the month. *(8L–7M)*

(c) Describe two factors which could account for the sharpness of peak flow at both gauging stations. *(9L–8M)*

(d) What is the most likely reason for the regular pattern of minor peaks recorded at station 99 on the River Hogsmill? *(5L–6M)*

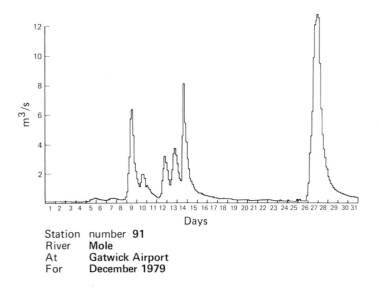

Station number **91**
River **Mole**
At **Gatwick Airport**
For **December 1979**

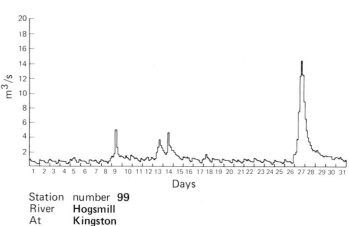

Station number **99**
River **Hogsmill**
At **Kingston**
For **December 1979**

The two hydrographs show the rate of streamflow at gauging stations on two tributaries of the River Thames for December 1979

29

End of month Soil Moisture Deficit in mm: Areal average values for the Lambourn catchment
(a tributary of the R. Thames)

Water Year	O	N	D	J	F	M	A	M	J	J	A	S
1974–75	2	0	1	0	5	2	20	59	123	146	167	109
1975–76	118	73	52	51	41	59	111	144	172	202	223	181
1976–77	96	34	0	0	0	1	19	64	63	113	31	59
1977–78	33	1	0	0	0	0	4	54	106	83	112	127
1978–79	135	109	0	0	0	2	16	3	51	102	116	127
1979–80	100	56	1	0	4	0	54	103	104	124	103	89
1980–81	27	0	0	2	0	0	7	11	81	84	116	30
MEAN	64	34	7	7	7	9	33	63	100	122	124	103

(a) What is the Soil Moisture Deficit? *(4L–3M)*

(b) Account for the major seasonal variations shown in the table above. *(8L–9M)*

(c) Suggest reasons for the anomaly of 1975–76 for the winter half of the year. *(6L–6M)*

(d) To whom would such information be of major interest and importance? *(6L–7M)*

30

(a) Define (i) interception loss, (ii) stemflow, (iii) throughfall. *(3 × [3L–3M])*

(b) At what stage and why in a storm will interception loss be greatest? *(5L–4M)*

(c) What are the major factors that govern the rate of interception loss? *(8L–7M)*

(d) Describe how either (i) interception loss or (ii) stemflow could be measured in the field. *(6L–5M)*

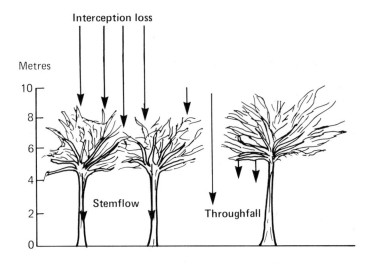

Precipitation avenues in a wooded area

31

(a) What is the average maximum length of overland flow in (i) basin A? (ii) basin B? *(2L–4M)*

(b) Given uniformity in all other factors, what effect does the length of overland flow have on discharge? *(10L–9M)*

(c) Explain three other important factors which could affect the rate of discharge in a drainage basin. *(12L–12M)*

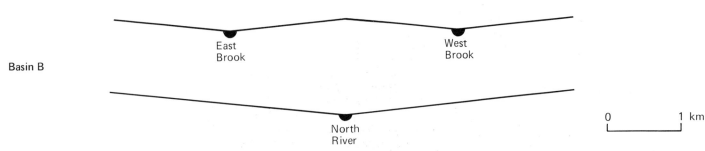

32

(a) Account for the shape of the infiltration curve shown on the diagram. *(5L–6M)*

(b) On a copy of the diagram insert a possible infiltration curve for a heavily grazed soil in an adjacent area with the same precipitation. *(5M)*

(c) Outline the other factors which can affect infiltration capacity. *(10L–8M)*

(d) Describe a method that can be used in the field to measure the infiltration capacity of a soil. *(6L–6M)*

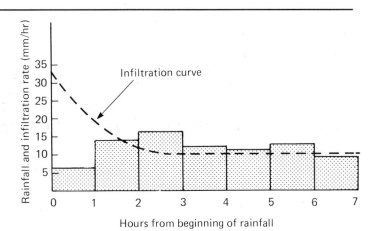

Infiltration capacity of a normal soil

33

Report on the Heavy Rainfall and Flooding of Late December

The first half of December was wet, with some heavy rainfall between the 9th and 14th. There was then little rain until the 26th.

In the last week of December, the British Isles were affected by the fronts of a slow moving depression centred over Iceland. Two associated depressions to the south west of Britain diverged, one moving north-east over Ireland and Scotland and the other moving east over northern France. The complex series of fronts between these two low pressure centres moved east across Britain on 27 December, bringing heavy rain to most of the country and causing severe flooding in many areas.

The fronts brought generally continuous rainfall to the whole of the Thames catchment between 03.00 on 27 and 05.00 on 28 December. Over 50 mm of rain were recorded on the Mole and Wey, the upper Kennet, the upper Colne and the northwestern edge of the catchment from the Churn to the upper Cherwell. Falls of about 25 mm were recorded in the Thame, lower Kennet and Loddon catchments.

This caused flooding in many of the larger tributaries, but the duration of the rainfall was not great enough to cause problems on the lower Thames. (Thames Water Authority)

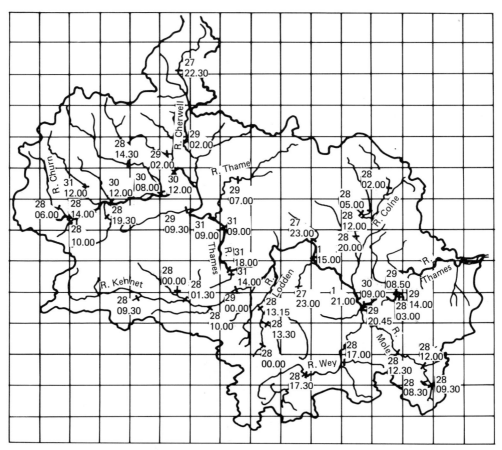

Dates and estimated times of flood peaks at gauging stations, December–January 1979–80

(a) Suggest why precipitation varied to such a degree over the Thames catchment. (6L–6M)

(b) What is a gauging station? (4L–3M)

(c) Suggest reasons for the marked time differences in flood peaks along the tributaries of the Thames. (7L–8M)

(d) Account for the different timings of flood peaks along the Thames itself. (7L–8M)

Hydrology – Section B

34

(a) Explain the relationship between precipitation and streamflow for the two rivers. *(6M)*

(b) What is the main reason for the variations in streamflow recorded at the four gauging stations along the River Mole? *(5M)*

(c) What was the time interval between each measurement of streamflow? *(2M)*

(d) Suggest reasons for the sharper peak flow and lower base flow of the River Mole. *(7M)*

(e) Account for the fact that some other tributaries of the Thames, such as the upper Kennet, which received some of the heaviest rainfall at this time, recorded no exceptional flows. *(5M)*

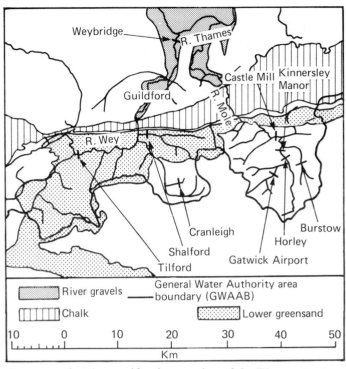

Outcrops of major aquifers in a section of the Thames basin

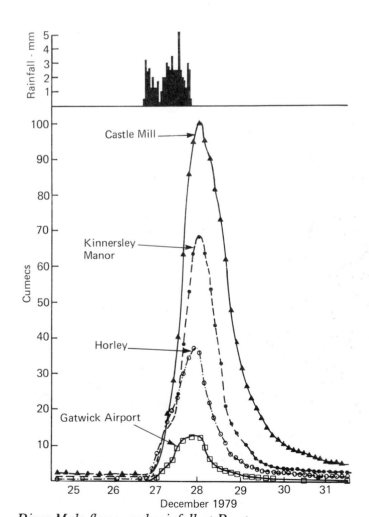

River Mole flows, and rainfall at Burstow

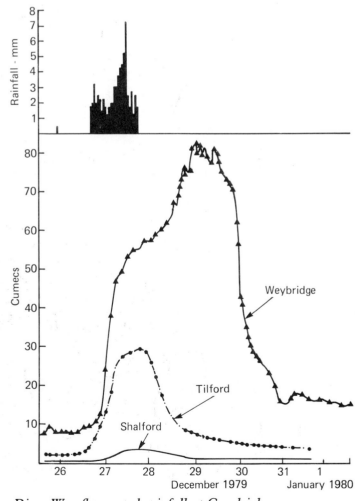

River Wey flows, and rainfall at Cranleigh

35

(a) Describe the path of the low pressure system. *(4M)*

(b) How can such a movement lead to a storm surge moving steadily southwards in the North Sea? *(7M)*

(c) Suggest the hydrological conditions in the Thames valley which could combine with such a storm surge to result in the flooding of the River Thames. *(6M)*

(d) Account for the difference between east and west London in the maximum area likely to be flooded. *(3M)*

(e) What is the rationale behind the Thames flood barrier? *(5M)*

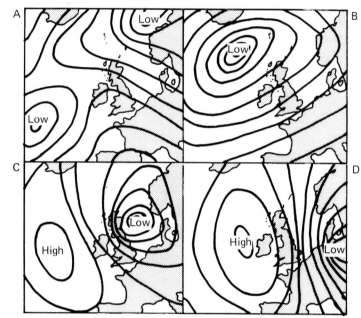

Figure 1

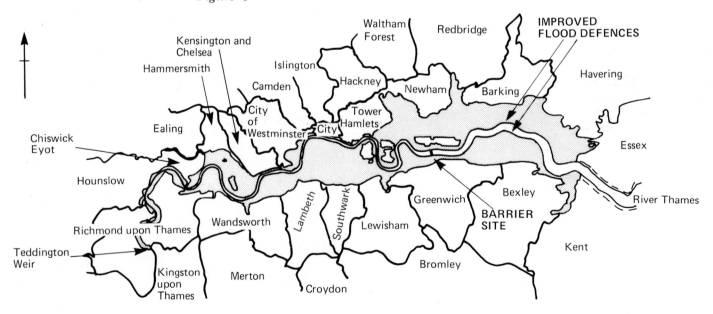

Figure 2   The maximum area likely to be affected by a surge tide flood

36

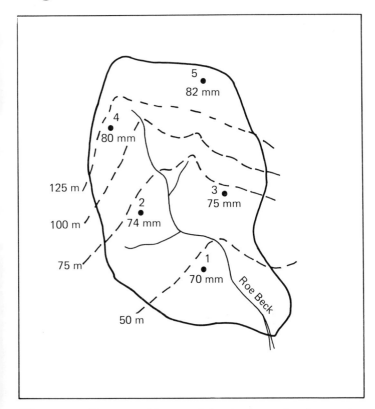

Figure 1   One month's precipitation at the five rain gauges in the catchment area

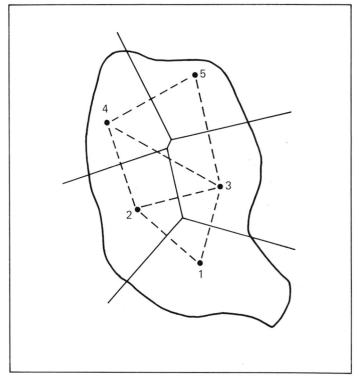

Figure 2   Thiessen polygons

Rain gauges	Precipitation mm	% Area of catchment in polygon	% Area of catchment between isohyets
1	70	30	32
2	74	20 }	29
3	75	17.5 }	
4	80	17.5	26.5
5	82	15	12.5

(a) Using the data presented above, calculate the average precipitation of the catchment area of Roe Beck by (i) the Thiessen polygon method and (ii) the Isohyetal method.   *(9M)*

(b) Why do hydrologists need to assess the average precipitation in a catchment area?   *(6M)*

(c) What precautions have to be taken in siting a rain gauge?   *(4M)*

(d) How can a rain gauge be adapted to record short interval (e.g. hourly) measurements?   *(3M)*

(e) If precipitation falls as snow how can its water equivalent be assessed?   *(3M)*

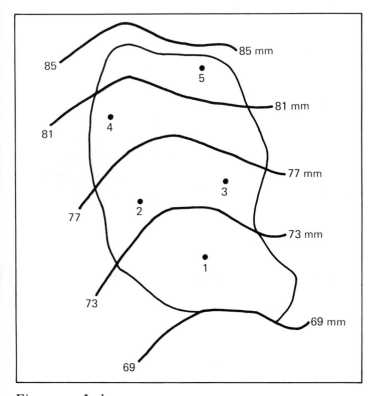

Figure 3   Isohyets

37

1980/81 Water Year	O	N	D	J	F	M	A	M	J	J	A	S	Total
Precipitation	97.5	63.0	56.3	25.6	30.3	167.0	52.3	104.1	31.8	73.4	40.3	135.6	877.1
Actual evaporation	22.1	5.8	3.2	5.1	12.6	29.5	51.6	82.1	98.9	68.7	66.7	31.9	478.1
Percolation	12.8	30.9	52.9	16.6	8.4	130.1	8.0	25.9	3.0	8.1	4.9	17.7	319.1
Soil moisture deficit	27	0	0	2	0	0	7	11	81	84	116	30	–

Tributary 1

1980/81 Water Year	O	N	D	J	F	M	A	M	J	J	A	S	Total
Precipitation	86.9	54.4	50.9	35.9	32.6	132.4	31.9	111.8	33.7	49.0	33.8	123.9	777.3
Actual evaporation	21.2	5.6	3.0	5.0	12.1	28.3	48.8	78.8	63.0	48.4	53.6	31.2	399.0
Percolation	54.2	47.6	47.7	31.9	15.1	120.9	3.2	23.9	3.4	4.1	3.8	21.9	377.8
Soil moisture deficit	2	0	0	1	2	0	20	11	44	47	71	0	–

Tributary 2

The hydrological data above was recorded in the catchment areas of two tributaries of the Thames.

(a) Define (i) Actual evaporation, (ii) Percolation, (iii) Soil moisture deficit. (6M)

(b) Comment on the seasonal variations in precipitation and actual evaporation for the two locations. (6M)

(c) Suggest reasons for the differences in percolation and soil moisture deficit shown in the tables. (8M)

(d) In which location, and why, might you expect streamflow to be most responsive to precipitation? (5M)

Meteorology and Climatology – Section A

38

(a) Name (i) Layer A, (ii) Layer B, (iii) Layer C, (iv) Layer D. (*4L–4M*)

(b) What are the correct meteorological terms applied to (i) a decline in temperature with altitude? (ii) an increase in temperature with altitude? (iii) no change in temperature with altitude? (*3L–3M*)

(c) Account for the temperature decreases in layers A and C. (*8L–9M*)

(d) Explain the temperature increases in layers B and D. (*8L–9M*)

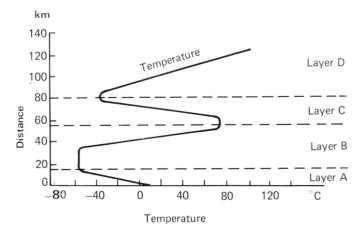

Thermal stratification of the atmosphere

39

(a) What percentage of solar radiation received at the top of the atmosphere is absorbed by the earth's surface? (*1L–2M*)

(b) How do solar and terrestial radiation differ? (*6L–5M*)

(c) Comment on the latitudinal variations in reflection from and absorption by the earth's surface. (*10L–10M*)

(d) Why is there such a latitudinal difference between solar radiation absorbed by cloud and radiation reflected by cloud? (*8L–8M*)

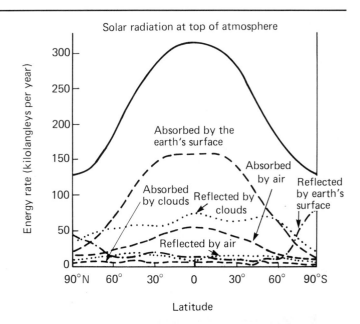

The average annual latitudinal disposition of solar radiation

27

40

(a) What is the approximate latitudinal divide between heat sources and heat sinks? *(1L–2M)*

(b) What percentage of heat transfer is carried by (i) ocean currents? (ii) atmospheric motion? *(2L–4M)*

(c) What is meant by (i) latent heat? (ii) sensible heat? *(2 × [3L–3M])*

(d) Explain and exemplify latitudinal heat transfer by atmospheric motion. *(15L–13M)*

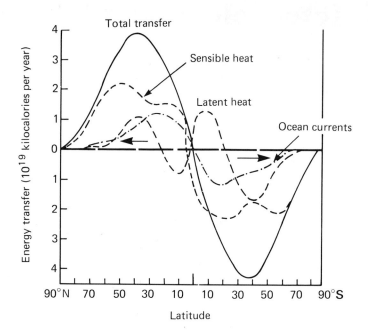

Heat transfer in the earth-atmosphere energy system

41

Study the January world isothermal map.

(a) On a copy of the map clearly mark and name two warm ocean currents and two cold ocean currents. *(4M)*

(b) What is the thermal equator? *(3L–3M)*

(c) Explain its position as shown on the map. *(8L–6M)*

(d) Account for the temperature variations along latitude 50°N. *(14L–12M)*

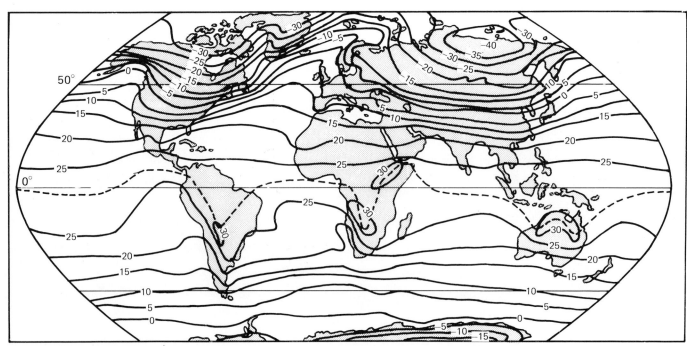

Mean sea level temperatures in January (the thermal equator is shown by the dashed line)

(a) What is the relative humidity when the following readings are obtained from the wet and dry bulb thermometers?

  (i) Dry bulb 32°C, wet bulb 25°C

  (ii) Dry bulb 20°C, wet bulb 18°C

  (iii) Dry bulb 12°C, wet bulb 8°C   *(3L–6M)*

(b) Explain why different temperature readings may be obtained from the wet and dry bulb thermometers.   *(11L–9M)*

(c) Why are relative humidity readings of such importance to the meteorologist?   *(12L–10M)*

Depression of wet bulb Degrees Centigrade

Dry Bulb temperature Degrees Centigrade / Relative humidity per cent

T	0.5	1	1.5	2	2.5	3	3.5	4	4.5	5	5.5	6	6.5	7	7.5	8	8.5	9	9.5	10	10.5	11	11.5	12	12.5	13	13.5	14	14.5	15	15.5	16	16.5	17
55	97	95	92	90	87	85	83	81	78	76	74	72	70	68	66	64	62	60	58	56	54	53	51	49	48	46	44	43	41	40	38	37	35	34
54	97	95	92	90	87	85	82	80	78	76	73	71	69	67	65	63	61	59	57	55	54	52	50	48	47	45	43	42	40	39	37	36	34	33
53	97	95	92	90	87	85	82	80	78	76	73	71	69	67	65	63	61	59	57	55	53	52	50	48	46	45	43	41	40	38	37	35	34	32
52	97	95	92	89	87	84	82	80	77	75	73	71	68	66	64	62	60	58	56	54	53	51	49	47	45	44	42	40	39	37	36	34	32	31
51	97	95	92	89	87	84	82	80	77	75	73	71	68	66	64	62	60	58	56	54	52	51	49	47	45	43	42	40	38	37	35	34	32	31
50	97	94	92	89	87	84	82	79	77	74	72	70	68	66	63	61	59	57	55	53	51	50	48	46	44	42	41	39	37	36	34	33	31	30
49	97	94	92	89	86	84	81	79	77	74	72	70	67	65	63	61	59	57	55	53	51	49	47	45	44	42	40	38	37	35	34	32	30	29
48	97	94	92	89	86	84	81	79	76	74	71	69	67	65	62	60	58	56	54	52	50	48	46	45	43	41	39	38	36	34	33	31	30	28
47	97	94	92	89	86	83	81	78	76	73	71	69	66	64	62	60	58	56	54	52	50	48	46	44	42	40	39	37	35	34	32	30	29	27
46	97	94	91	89	86	83	81	78	76	73	71	68	66	64	62	59	57	55	53	51	49	47	45	43	41	40	38	36	34	33	31	29	28	26
45	97	94	91	88	86	83	80	78	75	73	70	68	66	63	61	69	57	54	52	50	48	46	44	42	41	39	37	35	33	32	30	28	27	25
44	97	94	91	88	86	83	80	78	75	72	70	68	65	63	61	58	56	54	52	50	48	46	44	42	40	38	36	34	33	31	29	27	26	24
43	97	94	91	88	85	83	80	77	75	72	70	67	65	62	60	58	55	53	51	49	47	45	43	41	39	37	35	33	32	30	28	26	25	23
42	97	94	91	88	85	82	80	77	74	72	69	67	64	62	59	57	55	53	50	48	46	44	42	40	38	36	34	32	31	29	27	25	24	22
41	97	94	91	88	85	82	79	77	74	71	69	66	64	61	59	56	54	52	50	47	45	43	41	39	37	35	33	31	30	28	26	24	23	21
40	97	94	91	88	85	82	79	76	73	71	68	66	63	61	58	56	53	51	49	47	45	42	40	38	36	34	32	30	29	27	25	23	21	20
39	97	94	91	87	84	82	79	76	73	70	68	65	63	60	58	55	53	50	48	46	44	41	39	37	35	33	31	29	27	26	24	22	20	18
38	97	94	90	87	84	81	78	75	73	70	67	65	62	59	57	54	52	50	47	45	43	41	38	36	34	32	30	28	26	24	22	21	19	17
37	97	93	90	87	84	81	78	75	72	69	67	64	61	59	56	54	51	49	46	44	42	40	37	35	33	31	29	27	25	23	21	19	17	16
36	97	93	90	87	84	81	78	75	72	69	66	63	61	58	55	53	50	48	45	43	41	39	36	34	32	30	28	26	24	22	20	18	16	14
35	97	93	90	87	83	80	77	74	71	68	65	63	60	57	55	52	49	47	45	42	40	37	35	33	31	29	26	24	22	20	18	16	14	13
34	96	93	90	86	83	80	77	74	71	68	65	62	59	56	54	51	49	46	44	41	39	36	34	32	29	27	25	23	21	19	17	15	13	11
33	96	93	89	86	83	80	76	73	70	67	64	61	58	56	53	50	48	45	42	40	37	35	33	30	28	26	24	21	19	17	15	13	11	9
32	96	93	89	86	83	79	76	73	70	67	64	61	58	55	52	49	47	44	41	39	36	34	31	29	27	24	22	20	18	16	13	11	9	7
31	96	93	89	86	82	79	75	72	69	66	63	60	57	54	51	48	46	43	40	38	35	32	30	28	25	24	20	18	16	14	12	9	7	5
30	96	93	89	85	82	78	75	72	68	65	62	59	56	53	50	47	44	42	39	36	34	31	29	26	24	21	19	16	14	12	10	8	5	3
29	96	92	89	85	81	78	74	71	68	65	61	58	55	52	49	46	43	40	38	35	32	30	27	24	22	19	17	15	12	10	8	5	3	1
28	96	92	88	85	81	77	74	70	67	64	60	57	54	51	48	45	42	39	36	33	31	28	25	23	20	18	15	13	10	8	6	3	1	
27	96	92	88	84	81	77	73	70	66	63	60	56	53	50	47	44	41	38	35	32	29	26	24	21	18	16	13	11	8	6	3	1		
26	96	92	88	84	80	76	73	69	66	62	59	55	52	49	46	42	39	33	30	27	25	22	19	16	14	11	8	6	3	1				
25	96	92	88	84	80	76	72	68	65	61	58	54	51	47	44	41	38	35	32	29	26	23	20	17	14	11	9	6	3	1				
24	96	91	87	83	79	75	71	68	64	60	57	53	50	46	43	39	36	33	30	27	24	21	18	15	12	9	6	4	1					
23	96	91	87	83	79	75	71	67	63	59	56	52	48	45	41	38	35	31	28	25	22	19	16	13	10	7	4	1						
22	95	91	87	82	78	74	70	66	62	58	54	51	47	43	40	36	33	29	26	23	20	16	13	10	7	4	1							
21	95	91	86	82	78	73	69	65	61	57	53	49	45	42	38	35	31	27	24	21	17	14	11	8	4	1								
20	95	91	86	81	77	73	68	64	60	56	52	48	44	40	36	33	29	25	22	18	15	11	8	5	2									
19	95	90	86	81	76	72	67	63	59	55	50	46	42	38	34	31	27	23	19	16	12	9	5	2										
18	95	90	85	80	76	71	66	62	58	53	49	45	41	36	32	29	25	21	17	13	10	6	2											
17	95	90	85	80	75	70	65	61	56	52	47	43	39	34	30	26	22	18	14	10	7	3												
16	95	89	84	79	74	69	64	60	55	50	46	41	37	32	28	24	20	16	11	7	4													
15	94	89	84	78	73	68	63	58	53	49	44	39	35	30	26	21	17	13	8	4														
14	94	89	83	78	72	67	62	57	52	47	42	37	32	28	23	18	14	10	5	1														
13	94	88	83	77	71	66	61	55	50	45	40	35	30	25	20	16	11	6	2															
12	94	88	82	76	70	65	59	54	48	43	38	32	27	22	17	12	8	3																
11	94	87	81	75	69	63	58	52	46	41	35	30	25	19	14	9	4																	
10	93	87	81	74	68	62	56	50	44	38	33	27	22	16	11	5																		
9	93	86	80	73	67	61	54	48	42	36	30	24	18	13	7	2																		
8	93	86	79	72	66	59	52	46	40	33	27	21	15	9	3																			
7	93	85	78	71	64	57	50	44	37	31	24	18	11	5																				
6	92	85	77	70	63	55	49	41	34	28	21	14																						
5	92	84	76	69	61	53	46	39	31	24																								
4	92	83	75	67	59	51	44	36																										
3	91	83	74	66	57	49																												
2	91	82	73	64																														
1	90	81																																

43

Meteorological definitions

A: A mass of water droplets present in the lower layers of the atmosphere caused by the condensation of water vapour in the air reducing visibility to less than 2 km but not more than 1 km.

B: A dense mass of small water drops or smoke or dust particles in the lower layers of the atmosphere reducing visibility to less than 1 km.

C: A mass of minute solid particles of dust, smoke etc., reducing visibility to less than 2 km but not more than 1 km.

(a) Name the phenomena described, (i)A, (ii)B, (iii)C.
(3L–6M)

(b) Describe the natural atmospheric conditions which favour the formation of one of the features named in (a).
(12L–12M)

(c) What effects can human geography have on any of the phenomena identified?
(9L–7M)

44

(a) Suggest the approximate annual precipitation (in mm) at (i) place A, (ii) place B.
(2L–4M)

(b) Name the likely dominant type of rainfall in the region covered by the diagram.
(1L–2M)

(c) Explain the precipitation variations and the location of forested slopes shown on the diagram.
(12L–12M)

(d) Describe one other possible contributory type of rainfall affecting the region.
(9L–7M)

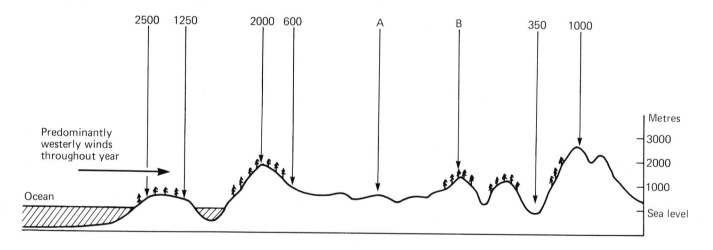

↟ ↟ ↟ Forested slopes

Annual precipitation (mm)

45

(a) Outline the general conditions necessary for the formation of dew. *(10L–10M)*

(b) Why is the intensity of dew formation usually greater on vegetation than on bare soil? *(10L–10M)*

(c) Describe the linked phenomenon that is likely to occur when the temperature falls below 0°C. *(5L–5M)*

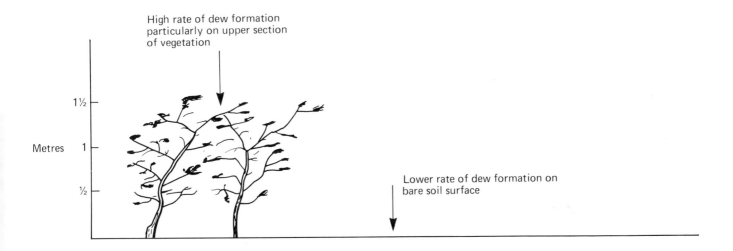

Dew formation on vegetation and soil surfaces

46

(a) On a copy of the diagram insert a likely position for the 0°C isotherm. *(4M)*

(b) Explain the relationship between snowfall and altitude. *(8L–6M)*

(c) Why is drifting more likely in upland rather than lowland areas? *(8L–8M)*

(d) What is the snowline and why does it vary with latitude? *(10L–7M)*

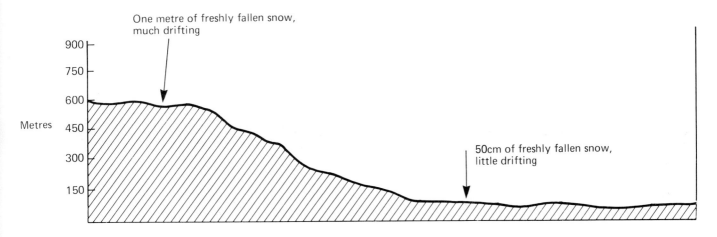

Variations in snowfall with altitude

47

(a) What are the units of pressure shown on the diagram? (*1L–2M*)

(b) What are the lines joining equal pressure areas called? (*1L–2M*)

(c) Name (i) Force A, (ii) Force B, (iii) Wind C. (*3L–3M*)

(d) Explain (i) Force A, (ii) Force B, (iii) Wind C. (*3 × [8L–6M]*)

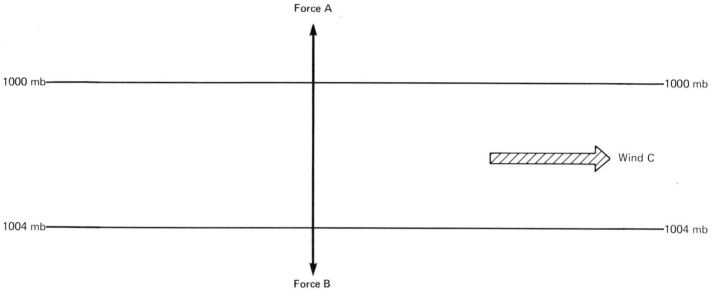

The forces governing atmospheric motion

48

(a) What are the alternative meteorological terms given to valley and mountain winds? (*1L–2M*)

(b) Account for the isothermal pattern shown in diagram A. (*6L–6M*)

(c) On a copy of diagram B insert an isothermal pattern that might result in the airflow indicated. (*6M*)

(d) Describe the general characteristics of valley and mountain winds. (*14L–11M*)

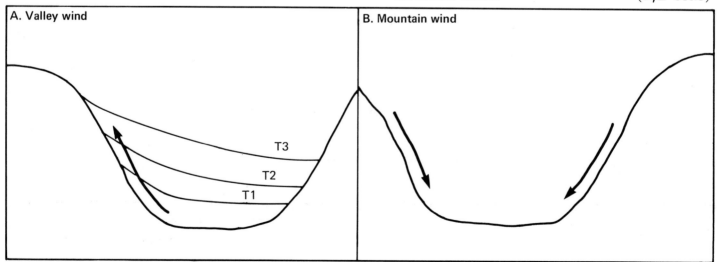

T1 = Highest temperature

49

(a) Name (i) Wind A, (ii) Wind B.   (2L–4M)

(b) On a copy of diagram 2 insert an isobaric pattern which might result in Wind B.   (5M)

(c) Describe and explain the causes and characteristics of (i) Wind A, (ii) Wind B.
       (2 × [10L–8M])

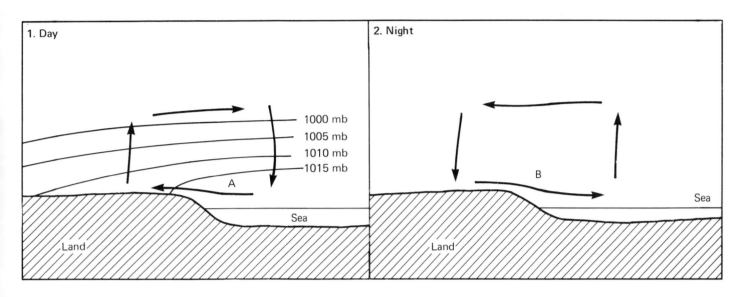

50

(a) On a copy of the diagram shade the area likely to be cloud covered.   (4M)

(b) Assuming a sea level temperature of 20°C what will the approximate temperature of the airstream be at (i) A, (ii) B, (iii) C, (iv) D?
       (4L–4M)

(c) Define (i) Condensation level, (ii) Dew-point temperature.   (2 × [2L–2M])

(d) What are the names given to such a phenomenon in (i) Europe? (ii) North America?   (2L–2M)

(e) Account for the temperature changes in the airstream as it passes over the mountain barrier.
       (13L–11M)

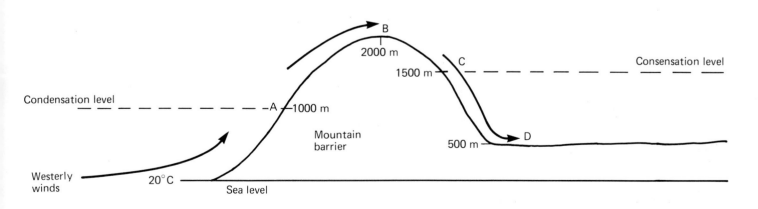

51

(a) Name the atmospheric processes operating at (i) A, (ii) B, (iii) C. (3L–3M)

(b) Name Wind D. (1L–1M)

SE Trade wind

Simplified diagram of inter-tropical air circulation

(c) On a copy of the diagram label (i) an area of high and non-seasonal rainfall, (ii) an area of very low annual rainfall, (iii) an area experiencing wet and dry seasons. (6M)

(d) Account for the main climatic similarities and differences between two of the areas identified in (c). (20L–15M)

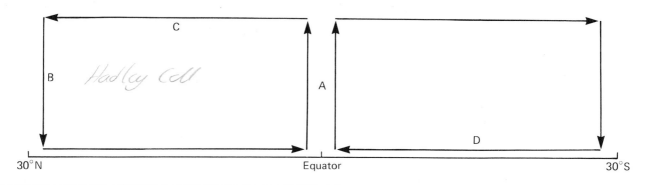

Hadley Cell

30°N   Equator   30°S

52

(a) Identify the major forces governing the wave pattern of mid-latitude upper air circulation. (10L–10M)

(b) In terms of the earth-atmosphere energy system, what is the main function of the mid-latitude waves? (6L–5M)

(c) Describe the relationship between the upper air circulation and surface anticylones and depressions. (10L–10M)

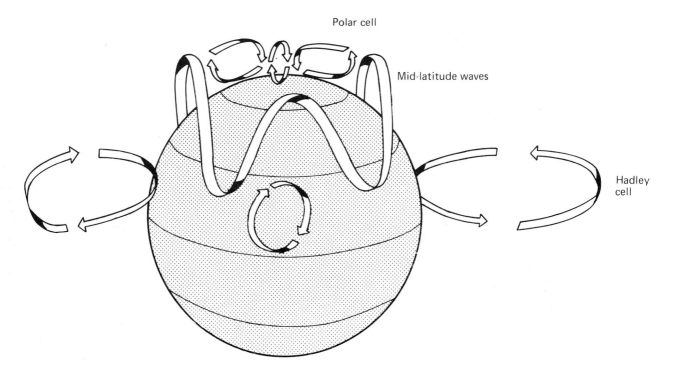

Polar cell

Mid-latitude waves

Hadley cell

53

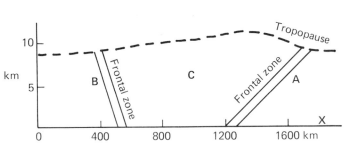

Direction of movement

(a) On a copy of the diagram shade and label the cloud types associated with such a system. *(6M)*

(b) Name (i) Front A, (ii) Front B, (iii) Area C. *(3L–3M)*

(c) Describe how the following might change at X as the system passes over: (i) Temperature, (ii) Pressure, (iii) Wind, (iv) Precipitation. *(4 × [5L–4M])*

Ana-frontal system

54

(a) On a copy of diagram B insert (i) the cold front, the warm front, the warm occlusion (ii) isotherms to show the vertical temperature variations. *(7M)*

(b) Show how an occluded front appears on a weather map. *(1L–2M)*

(c) Describe the causes and general weather conditions associated with occlusions. *(12L–10M)*

(d) Explain the differences between warm and cold occlusions. *(6L–6M)*

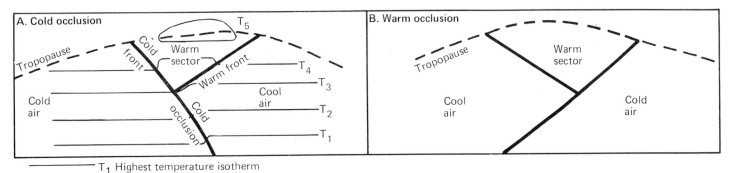

55

(a) Assess the maximum heat-island intensity illustrated by the diagram *(1L–2M)*

(b) Identify the factors responsible for such a temperature gradient. *(12L–11M)*

(c) Briefly explain four other characteristics of urban climates. *(4 × [4L–3M])*

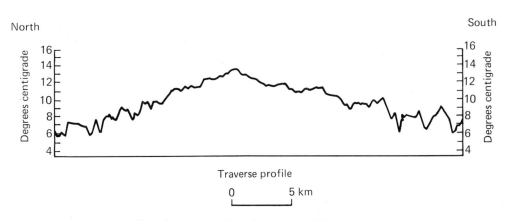

Temperature traverse, London 11–12 October 1961 (night)

(a) Compare the weather conditions at stations A and B. *(8L–8M)*

(b) Attempt to account for the main difference in weather between the two stations. *(8L–10M)*

(c) Identify and account for west to east temperature variations along latitude 60°N. *(8L–7M)*

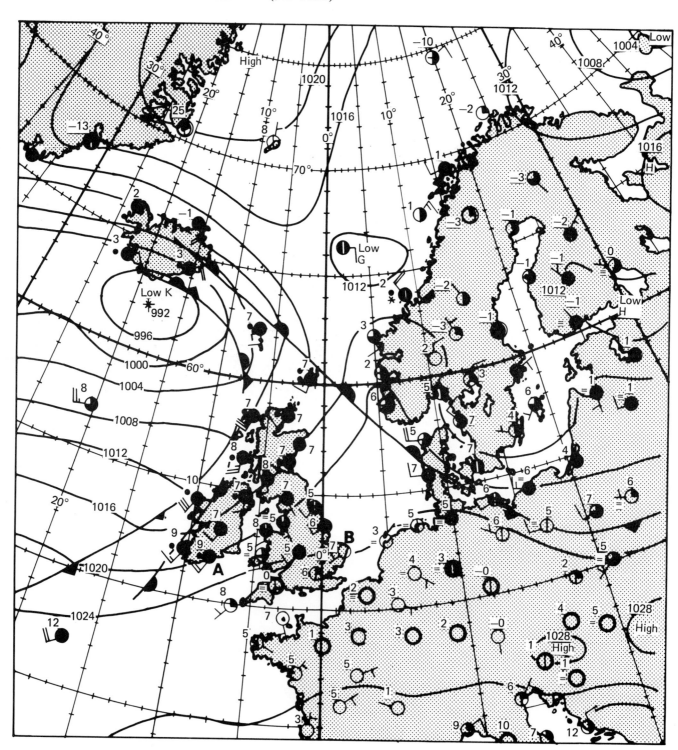

Western Europe, 0600h, 1 May 1976

57

(a) What does the continuity chart show? (8L–6M)

(b) Why are such charts of great importance to the meteorologist? (8L–8M)

(c) Explain how continuity charts are constructed. (9L–7M)

(d) To an observer on the west coast of Ireland, what would be the first visual signs of the approaching weather system? (4L–4M)

Continuity chart

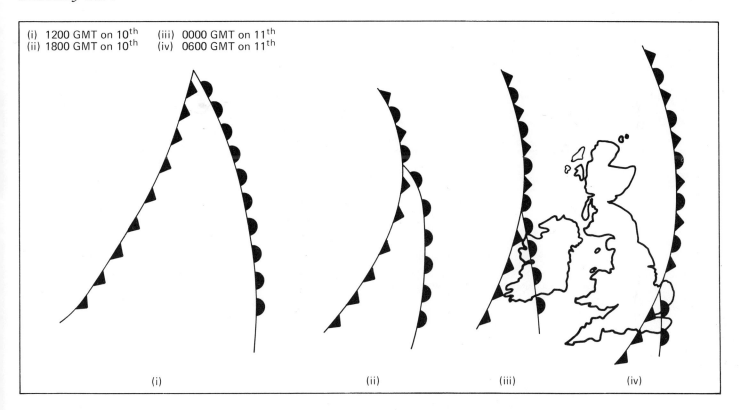

(i) 1200 GMT on 10th   (iii) 0000 GMT on 11th
(ii) 1800 GMT on 10th   (iv) 0600 GMT on 11th

(i)   (ii)   (iii)   (iv)

Meteorology and Climatology – Section B

58

(a) How is mean annual precipitation calculated? (*1M*)

(b) List three precipitation characteristics which the map does not illustrate. (*3M*)

(c) Why do high latitude regions have low precipitation? (*4M*)

(d) Briefly account for variations of precipitation within the tropics. (*7M*)

(e) What are the main factors governing precipitation totals in cool temperate latitudes? (*6M*)

(f) Assess the relationship between total precipitation and altitude. (*4M*)

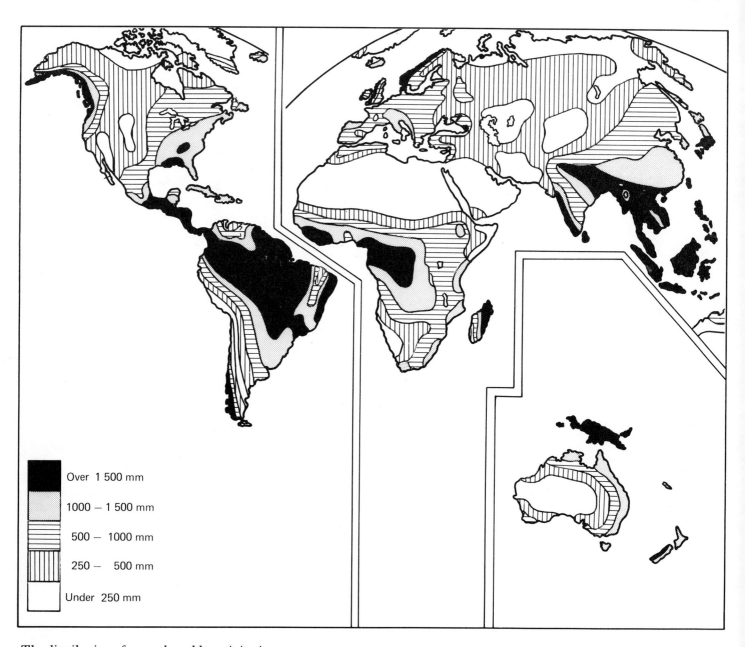

Over 1 500 mm

1000 – 1 500 mm

500 – 1000 mm

250 – 500 mm

Under 250 mm

The distribution of annual world precipitation

59

(a) At which place on the diagram will the largest hailstones fall, A, B, C or D? *(2M)*

(b) Explain the conditions under which thunderstorms are likely to develop. *(8M)*

(c) Account for the formation of clear and opaque layers of ice in large hailstones. *(8M)*

(d) Why is a well defined downdraught characteristic of thunderstorms? *(4M)*

(e) Account for the anvil shape in the upper part of the cloud. *(3M)*

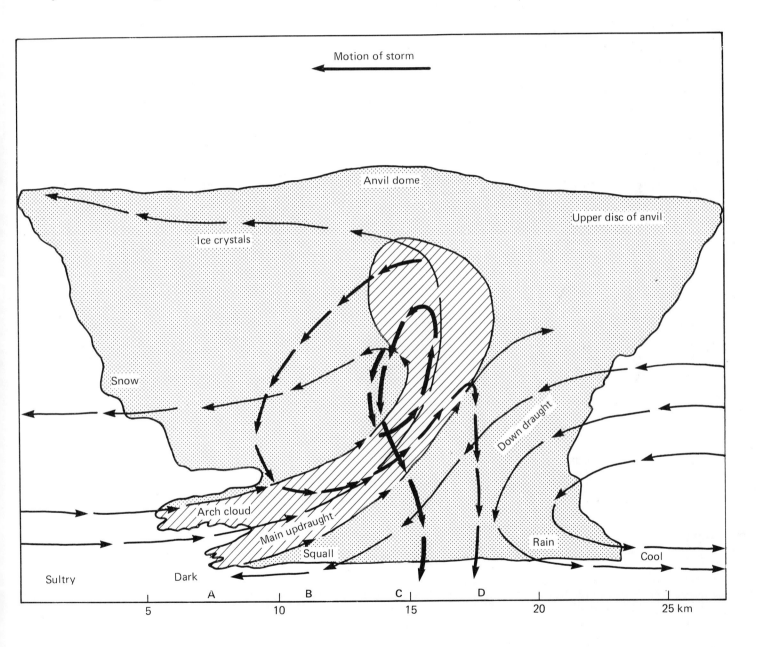

60

(a) On a copy of the graph provided, plot the path of a rising air parcel using the following information: (i) Ground temperature 10°C; (ii) From the surface to 500 m the air rises and cools at the DALR (10°C/1000 m). The dew-point is reached at 500 m; (iii) Above 500 m the air rises at the SALR (6°C/1000 m)* until the temperature of the rising air equals that of its surroundings. (4M)

(b) From the completed diagram state (i) the dew-point temperature, (ii) the level of the cloud base, (iii) the level of the cloud top, (iv) the vertical extent of the cloud, (v) the ELR between the surface and 1000 m, between 1000 m and 2000 m and between 2000 m and 3000 m. (7M)

(c) Explain why rising air cools at a slower rate once it is saturated. (4M)

(d) Under what conditions does relative instability occur? (5M)

(e) Identify the factors which could lead to changes in the ELR during the course of the day. (5M)

* For simplicity the SALR is taken to be constant here.

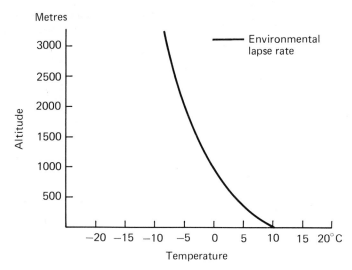

61

(a) Assess the annual range of temperature at A, B and C. (3M)

(b) In which climatic zone is each settlement located? Briefly outline the reasons for your answers. (6M)

(c) Account for the seasonal contrasts shown in C. Suggest a location for this settlement. (7M)

(d) Name four other parts of the world which have a broadly similar climate. (2M)

(e) Outline the advantages and disadvantages of such a climate to the economy of the region in which C is located. (7M)

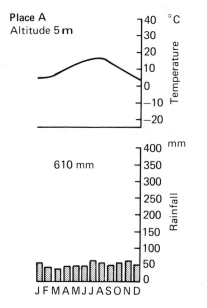

Place A
Altitude 5 m

610 mm

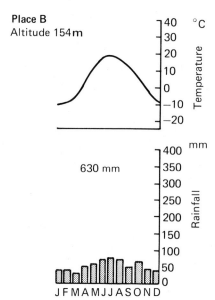

Place B
Altitude 154 m

630 mm

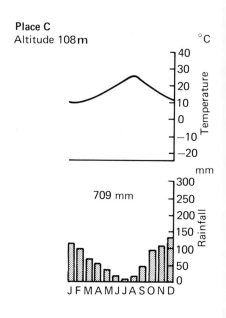

Place C
Altitude 108 m

709 mm

62

(a) Draw an annotated diagram to show the scale and major characteristics of a tropical hurricane.

(7M)

(b) Account for the source regions and the tracks of the major hurricanes to affect the USA as illustrated by the map.

(10M)

(c) Describe and explain which parts of the United States face the greatest danger from hurricanes.

(5M)

(d) Identify other world regions which are affected by similar phenomena.

(3M)

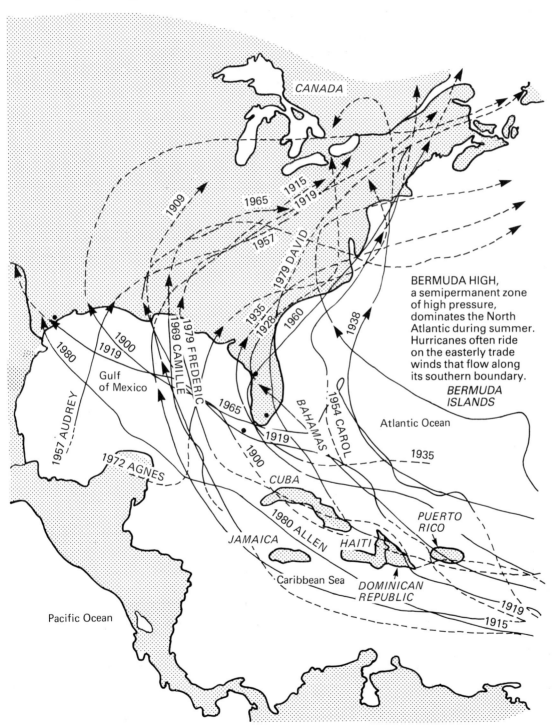

CANADA

1909
1965
1915
1919
1957
1935
1928
1960
1979 DAVID
1969 CAMILLE
1979 FREDERIC
1900
1919
1980
1957 AUDREY
Gulf of Mexico
1972 AGNES
1965
1919
1900
1965
1938
BAHAMAS
1954 CAROL

BERMUDA HIGH, a semipermanent zone of high pressure, dominates the North Atlantic during summer. Hurricanes often ride on the easterly trade winds that flow along its southern boundary.

BERMUDA ISLANDS

Atlantic Ocean

1935

PUERTO RICO

CUBA
1980 ALLEN
JAMAICA
HAITI
Caribbean Sea
DOMINICAN REPUBLIC
1919
1915

Pacific Ocean

Hurricanes on the US coast

Biogeography – Section A

63

(a) Name the missing factor. *(1L–2M)*

(b) Which of the five factors did Dokuchayev consider to be of the greatest importance in the determination of soil type? Give *two* reasons for your answer. *(5L–5M)*

(c) Briefly describe the influence of any *three* factors, besides the one named in (b), on the formation of soil. *(3 × [6L–6M])*

After V. Dokuchayev (USSR) 1846–1903

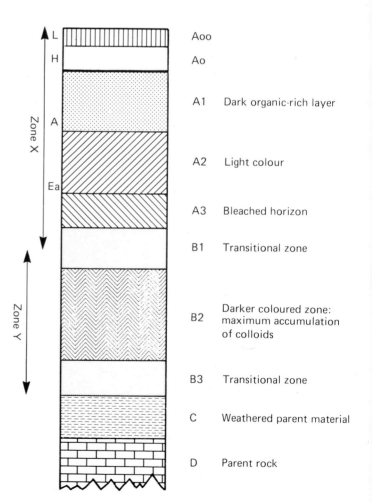

Climatic factor

Biotic factor

Soil

Geomorphic factor

Geological factor

64

(a) Name the Aoo and Ao horizons. Which factors affect the depth of these two layers? *(6L–8M)*

(b) Name the two soil drainage zones X and Y. *(2L–4M)*

(c) Explain how the colouring of the A and B horizons is related to the movement of water in these zones. *(8L–8M)*

(d) State why you think that the profile shown has developed under temperate conditions, with an excess of precipitation over evaporation. *(4L–5M)*

Aoo

Ao

A1   Dark organic-rich layer

A2   Light colour

A3   Bleached horizon

B1   Transitional zone

B2   Darker coloured zone: maximum accumulation of colloids

B3   Transitional zone

C   Weathered parent material

D   Parent rock

Zone X

Zone Y

The soil profile

65

(a) Identify the four zonal soil profiles shown.
(4L–8M)

(b) State which of the profiles you would expect to find at the following latitudes: (i) 2°N, (ii) 40°N, (iii) 60°N, (iv) 70°N.
(4L–4M)

(c) Explain the variation in thickness of the litter/humus layer shown in the profiles.
(6L–7M)

(d) For any *one* of the soil types shown, explain how the movement of water through the soil aids the development of the soil profile.
(5L–6M)

1

L/H	Acid humus
A1	Brown silt loam : acid
A2	Blue-grey clay/loam
C	Permafrost

2

L/H/A1	Rapid weathering and decomposition
A2	Grey-brown clay
B1	Red clay (iron oxides)
B2	Red sandy clay
C	Parent rock

3

L/H	
A1	Black crumby loam rich in fauna
A2–B	Concentration of calcium carbonate
C	Parent rock

4

L	
H	Acid humus
A1	Dark-brown acid layer
A2	Ash-grey (leached)
B1	Iron-pan
B2	Yellow-brown layer
C	Parent rock

Zonal soils

66

(a) Account for the differences in soil drainage characteristics on the hillslope.
(5L–6M)

(b) Give two reasons why you would expect leaching to be present on the slope.
(2 × [2L–2M])

(c) Choose any *one* of soil characteristics 6, 7 and 8 and describe how it could be tested.
(6L–6M)

(d) What conclusions can be drawn from the results shown on the transect regarding the development of soil cover in the study area?
(8L–9M)

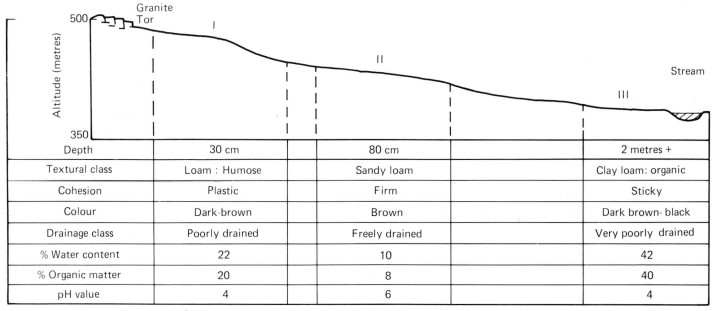

Depth	30 cm	80 cm	2 metres +
Textural class	Loam : Humose	Sandy loam	Clay loam: organic
Cohesion	Plastic	Firm	Sticky
Colour	Dark-brown	Brown	Dark brown- black
Drainage class	Poorly drained	Freely drained	Very poorly drained
% Water content	22	10	42
% Organic matter	20	8	40
pH value	4	6	4

A transect diagram to show the results of an A-level student's field investigation of soil characteristics on a Dartmoor hillslope

67

(a) What name is given to the regular succession of soil profiles down a slope? *(1L–2M)*

(b) Using profile 1 as a guide, draw the three other profiles shown on the diagram. Annotate the main horizons on each profile. *(3 × 5M)*

(c) Account for the variation in soil profiles down the slope. *(6L–8M)*

The succession of soil profiles down a slope

Relief	Upland plateau	Upper slope (20–35°)	Lower slope (5–15°)	Valley bottom
Height	Over 300m	150–300 m	50–150 m	Below 50 m
Land use	Moorland	Upland grazing	Arable	Meadow
Rainfall	Over 1500 mm	1000–1500 mm	750–1000 mm	500 mm

A1 — Peat

A2 — Grey-green horizon

Parent rock

C

Hill peat

1

Podsol

2

Brown earth

3

Gley

4

68

(a) Explain the essential difference between a Pedalfer and a Pedocal and how this difference can be related to climatic factors. *(8L–10M)*

(b) Rank the Pedocal soils in order of increasing temperature, (low – high). *(4L–4M)*

(c) Rank the Pedalfer soils in order of increasing aridity, (high rainfall – low rainfall). *(4L–4M)*

(d) Pedalfer soils tend to be either leached or lateritic; explain the difference between these two terms. *(6L–7M)*

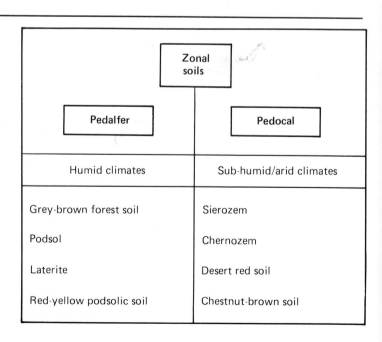

	Zonal soils	
Pedalfer		Pedocal
Humid climates		Sub-humid/arid climates
Grey-brown forest soil		Sierozem
Podsol		Chernozem
Laterite		Desert red soil
Red-yellow podsolic soil		Chestnut-brown soil

A classification of zonal soils

69

(a) Identify the five Global Ecosystems (Biomes) described in the table, suggesting a possible location in each case. (5L–10M)

(b) Define the terms Biomass and Net Primary Productivity and account for the great variation that exists in these two columns. (8L–8M)

(c) What other hydrological data besides Total Annual Precipitation is useful when studying Biomes? (5L–3M)

(d) Explain why B and D, two contrasting Biomes, both have such low levels of Net Primary Productivity. (6L–4M)

Biomes	Total annual precipitation (mm)	Approximate annual temp. range (°C)	Soil type	Mean plant biomass (kg/m^2)	Mean net primary productivity (g/m^2/day)
A	750	11	Laterite	4.0	1.9
B	250	40	Gley	0.7	0.4
C	2500	3	Ferralitic	44.0	5.5
D	120	20	Solonchak	0.7	0.2
E	500	40	Chernozem	1.6	1.4

Global ecosystems (biomes)

70

(a) Name the five trophic levels that make up the biological community. (5L–5M)

(b) Explain how energy is transferred throughout each of the trophic levels. (10L–10M)

(c) The biological community is sometimes sub-divided into Autotrophs, Heterotrophs and Decomposers. Explain how this can be related to the diagram shown. (5L–5M)

(d) Describe the components of the physical environment. (5L–5M)

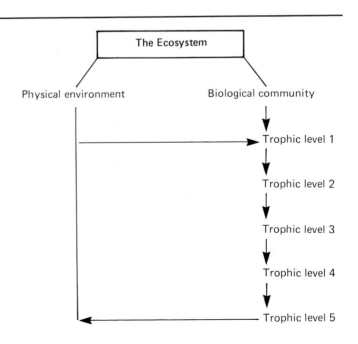

71

(a) Distinguish between Solar Radiation and Thermal Radiation. (4L–5M)

(b) Study the graph (Figure 2).
   (i) State the climax vegetation type at Station A. (1L–2M)

   (ii) Explain the use of net radiation at Station A, as shown on the graph. (6L–8M)

(c) Briefly describe the advantages and disadvantages of using vegetation cover as a measure of environmental energy. (8L–10M)

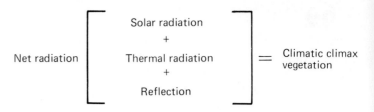

Figure 1   The energy budget equation

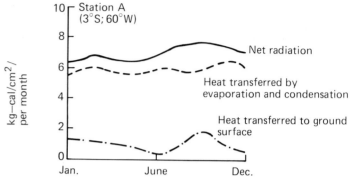

Figure 2

72

(a) What do you consider is the dominating factor in plant growth? (1L–1M)

(b) Name *four* finite resources which are vital to plant growth. (4L–4M)

(c) Describe the different methods by which (i) man and (ii) natural disasters disrupt the balance of the ecosystem. (2 × [4L–6M])

(d) Briefly explain the link shown on the diagram between soil and atmosphere. (3L–3M)

(e) Explain the ways in which plants adapt to climatic extremes. (5L–5M)

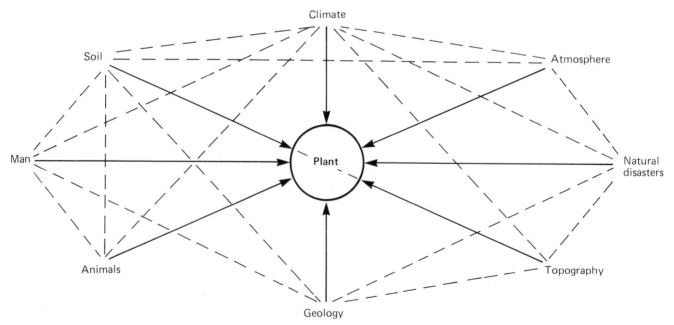

The environmental factors affecting plant growth

73

(a) Explain the essential difference between a seral community and a climatic climax community. *(4L–4M)*

(b) Draw a simple flow diagram to show the changes in vegetation which occur in the seral communities in the build up to climatic climax vegetation. *(2L–6M)*
(Any example of climax vegetation may be chosen.)

(c) Explain how man's activities can result in the formation of a plagioclimax community. *(6L–8M)*

(d) Why is it that the secondary succession in cycles 2 and 3 is often more rapid than the succession in cycle 1? *(5L–7M)*

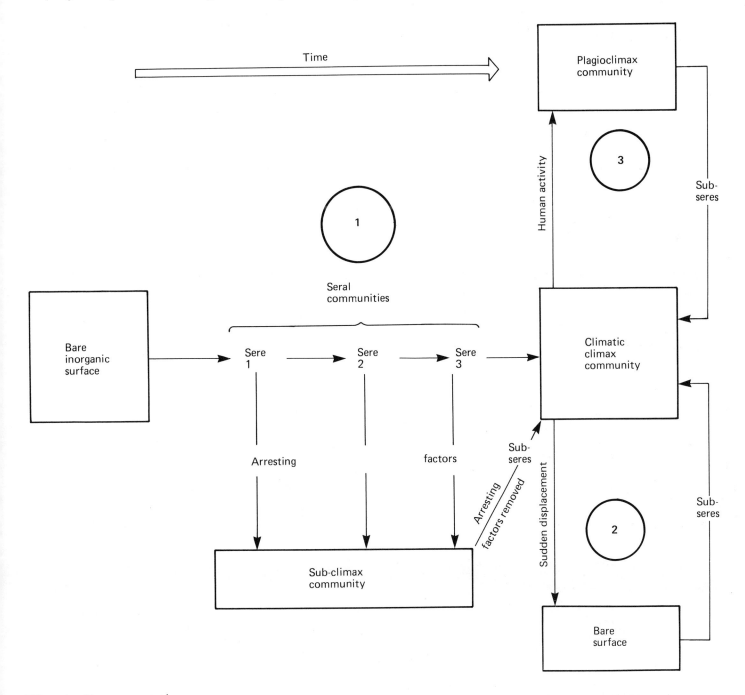

Climatic climax vegetation

74

(a) Define a prisere. *(2L–2M)*

(b) Explain the essential difference between a lithosere and a psammosere. Name and locate an example of each. *(6L–8M)*

(c) Describe the sequence of events that occurs in a plagiosere. *(6L–6M)*

(d) Study Figure 2 and draw an annotated diagram to show the final stage in the development of a hydrosere. *(9M)*

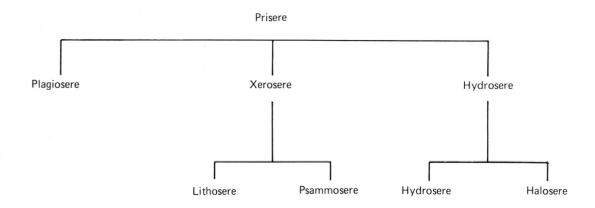

Figure 1

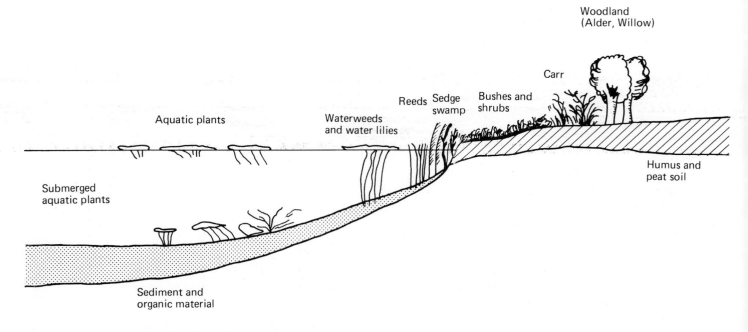

Figure 2 The development of a hydrosere, initial stage

75

(a) Name *one* example of each of the eight vegetation types listed in the tables. (8L–8M)

(b) Which combination of plants would you expect to find in (i) a Tundra environment? (ii) a Cool Temperate Deciduous Forest? (2 × [1L–2M])

(c) Explain how plants adapt to extreme rainfall conditions. (6L–6M)

(d) Account for the variation that exists in the tropical (megatherm) classification of vegetation. (6L–7M)

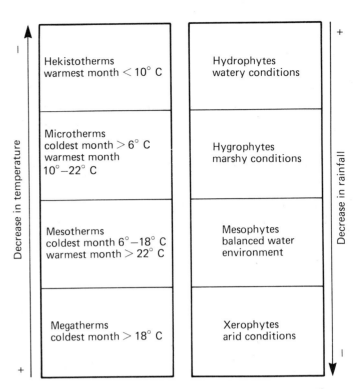

The influence of rainfall and temperature on vegetation

76

(a) The climatic regions shown by graphs A, B, C, D are as follows: Hot Desert, Arctic Tundra, Temperate and Tropical Rain Forest. State the correct letter for each region. (4L–8M)

(b) Explain the dominance of hemicryptophytes, therophytes and phanerophytes in *any three* of the climatic regions that you have named in (a). (3 × [4L–4M])

(c) Epiphytes and hydrophytes are not included in the broad classification by Raunkiaer. Define each of these terms and state in which of the four climatic regions they are likely to occur. (6L–5M)

Phanerophytes: (P)	Perennial shrubs and trees with stems and renewal buds more than 250 mm above the soil and exposed to most climatic hazards
Chamaephytes: (C)	Perennial herbs and low shrubs with renewal buds between ground level and 250 mm
Hemicryptophytes: (H)	Herbs and grasses with resting buds at ground level or in surface soil
Geophytes: (G)	Plants with underground bulbs, tubers or rhizomes well buried in the soil
Therophytes: (T)	Plants with a full life cycle under favourable conditions but which can survive stringent conditions in the form of resistant seeds or spores

A classification of vegetation, after C. Raunkiaer

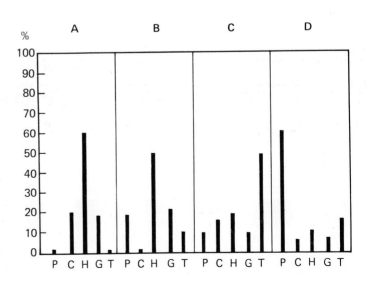

The Raunkiaer classification applied to broad climatic regions

(a) Briefly account for the changing distribution of vegetation along the line of the transect.

(6L–7M)

(b) State *four* reasons why plant communities in this type of environment are particularly resistant in their qualities of environmental adaptation.

(4 × [1L–2M]

(c) Describe the next stages in the plant succession of the psammosere. Why is it unlikely to attain climax vegetation? (6L–6M)

(d) Comment on impact of man on this coastal ecosystem. (4L–4M)

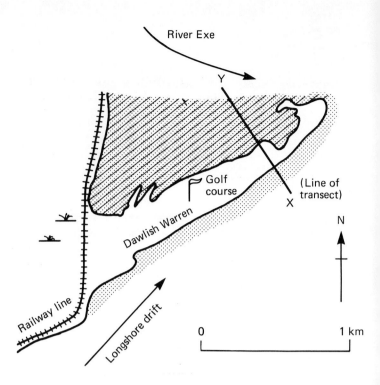

Dawlish Warren: A sand, shingle and salt-marsh environment

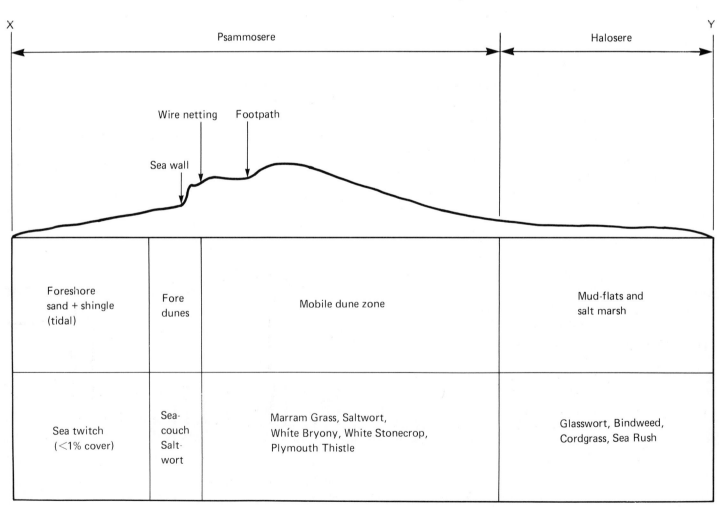

Foreshore sand + shingle (tidal)	Fore dunes	Mobile dune zone	Mud-flats and salt marsh
Sea twitch (<1% cover)	Sea-couch Salt-wort	Marram Grass, Saltwort, White Bryony, White Stonecrop, Plymouth Thistle	Glasswort, Bindweed, Cordgrass, Sea Rush

78

(a) Account for the variation in soil acidity shown on the diagram and explain the effect of this on the vegetation cover. (6L–7M)

(b) Draw an annotated flow diagram to represent the seral stages in the evolution of a chalk downland area in its attainment of climatic climax vegetation. (8M)

(c) Why can the sheeps fescue grass, which covers so much of the Chalk Downs, be regarded as plagioclimax vegetation? (4L–5M)

(d) Explain the role of geomorphic factors in the development of downland vegetation. (4L–5M)

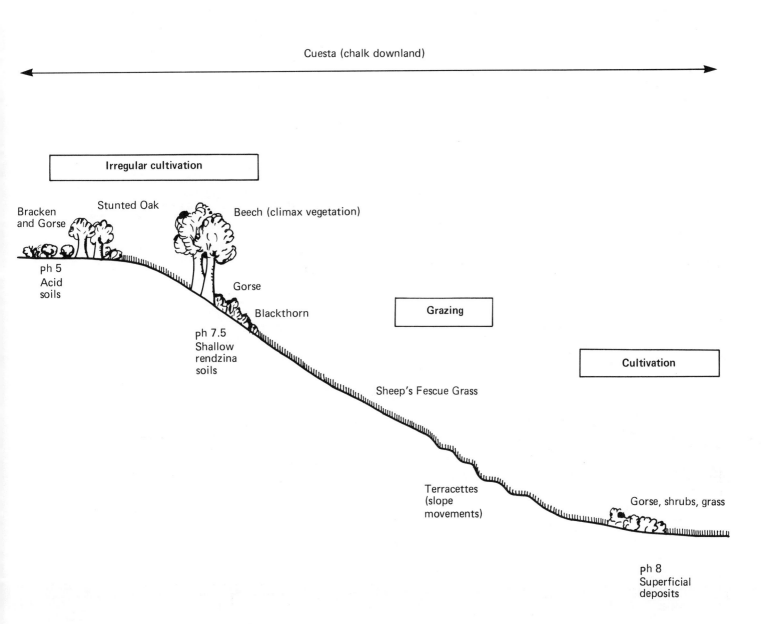

Vegetation and soil development on chalk downland

79

(a) Explain why the physical environment is ill-equipped to withstand any major disturbance of the ecosystem. (6L–6M)

(b) Which sections of the Food Chain will face the initial impact from mining developments? What will these immediate effects be? (4L–4M)

(c) Explain the role of invertebrates and micro-organisms in the Food Chain. (5L–7M)

(d) List some of the methods by which the mining companies can minimise their impact on (i) the animal population, (ii) the natural vegetation.
(2 × [4L–4M])

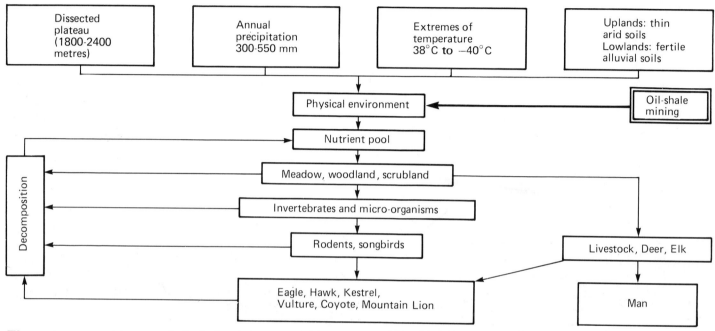

The environmental impact of oil-shale mining in the Rocky Mountains (USA)

80

(a) Describe the effects of each of the five negative environmental impacts shown on the diagram.
(5 × [3L–3M])

(b) In which areas are the effects of deforestation likely to be particularly severe? Use examples to illustrate your answer. (5L–5M)

(c) Describe the possible effects downstream of the hydrological changes mentioned in (a). (5L–5M)

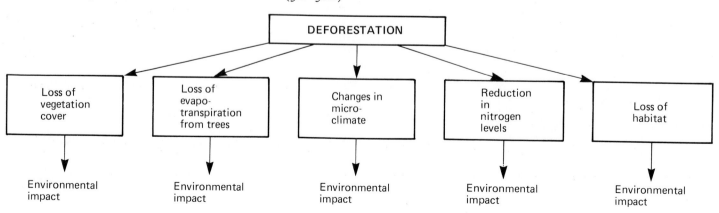

81

(a) Briefly account for the differences in vegetation between heathland and moorland. (6L–8M)

(b) The evidence of pollen analysis suggests that a climax vegetation of deciduous woodland existed in these regions approximately 7000 years ago. Using the information in Figure 4, explain the changes in vegetation that have taken place in these areas. (8L–8M)

(c) Although often termed waste land, moor and heath occupy one-third of Britain's land surface and should be seen as a valuable primary resource. Explain why. (9L–9M)

Figure 1   Heathland – The Ashdown Forest, Sussex

Figure 2   Moorland – Dartmoor

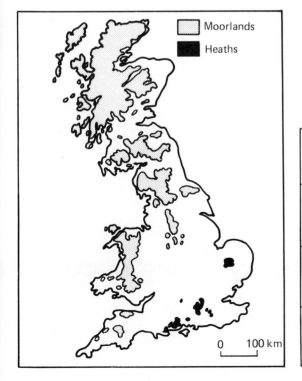

Figure 3

Figure 4

Years before present	7000 6000 5000 4000 3000 2000 1000
	Neolithic period Bronze age Iron age
Climatic conditions	Warm-wet (climatic optimum) Gradual deterioration → Cool-wet
Soils	Rich brown earths ——→ Acidic humus → Leaching → Podsol
Vegetation	Deciduous forest : Oak, Elm Oak-Birch forest (over 800m) → Pine Birch → Ling Heather Bracken → Moor Heath

53

82

(a) Describe the ways in which man has interfered with the hydrological balance in the National Park. (8L–6M)

(b) Describe other possible consequences which might result from man's increased use of the Park. (8L–8M)

(c) Suggest why the area shown in the photograph merits P1 status. (4L–6M)

(d) Suggest other means by which the environmental quality of the National Park might be preserved. (5L–5M)

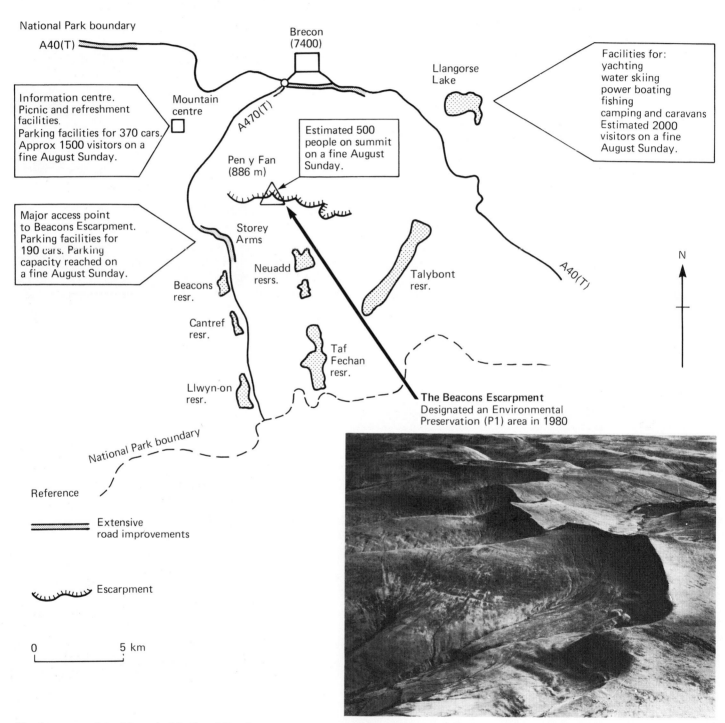

National Park boundary

A40(T)

Brecon (7400)

Llangorse Lake

Facilities for:
yachting
water skiing
power boating
fishing
camping and caravans
Estimated 2000 visitors on a fine August Sunday.

Information centre. Picnic and refreshment facilities. Parking facilities for 370 cars. Approx 1500 visitors on a fine August Sunday.

Mountain centre

A470(T)

Estimated 500 people on summit on a fine August Sunday.

Pen y Fan (886 m)

Major access point to Beacons Escarpment. Parking facilities for 190 cars. Parking capacity reached on a fine August Sunday.

Storey Arms

Neuadd resrs.

Talybont resr.

A40(T)

N

Beacons resr.

Cantref resr.

Taf Fechan resr.

Llwyn-on resr.

National Park boundary

The Beacons Escarpment
Designated an Environmental Preservation (P1) area in 1980

Reference

Extensive road improvements

Escarpment

0   5 km

Environmental problems in National Park areas – The Brecon Beacons

Biogeography – Section B

83

(a) Draw an annotated diagram similar to Figure 2 to show the transition in vegetation from the Equatorial Forest to the Hot Desert. Complete the annotation in the Equatorial Forest zone and label the other vegetation zones in the same fashion. (6M)

Climate and vegetation in West Africa

(b) Describe the relationship that exists between climate and vegetation in West Africa. (4M)

(c) Figure 3 shows the nutrient cycle of an Equatorial Rain Forest. (i) Explain how the nutrient cycle functions, (ii) Draw similar models to represent the Savanna and Hot Desert ecosystems. (3 × 5M)

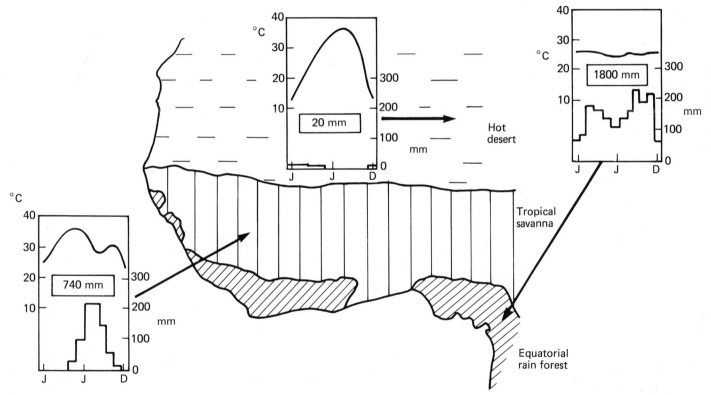

Figure 1

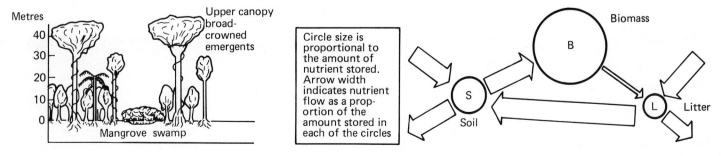

Figure 2

Figure 3 *The nutrient cycle of an equatorial rain forest*

The Norfolk Broads: A Freshwater Ecosystem under Pressure

'The problems facing the Norfolk Broads are acute and potentially disastrous. The balance of this fragile ecosystem is largely determined by the quality of its water supplies and in recent years, as a result of man's activity, the quality of this resource has deteriorated significantly.' (*Geographical Magazine*, October 1979)

(a) Explain why the Norfolk Broads can be described as a 'fragile ecosystem'. (*5M*)

(b) Describe the causes and effects of the eutrophication (nutrient enrichment) of the Broads. (*10M*)

(c) Assess the impact of the expanding tourist and leisure industry on the region. (*5M*)

(d) It is felt by some that by granting National Park status to the Norfolk Broads, overall control of land and water resources can be obtained. Suggest some possible solutions to the environmental management of the Norfolk Broads. (*5M*)

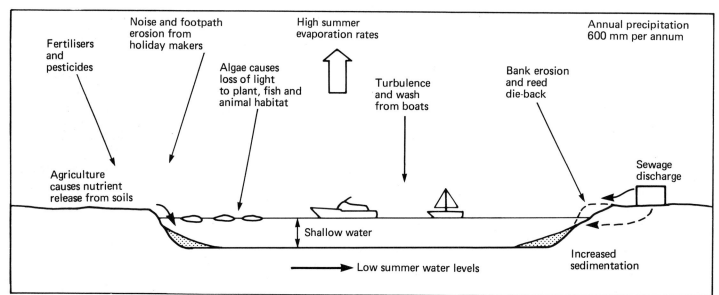

The nature of the problem

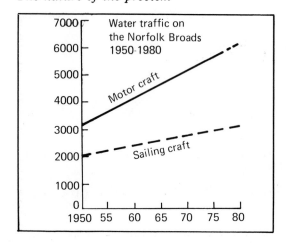

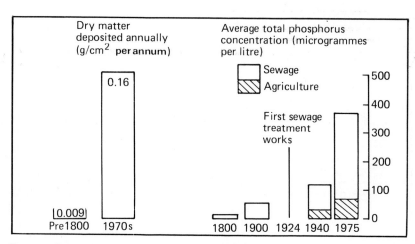

Barton Broad – Changes in water quality 1800–1975

85

(a) Draw and complete Table 1. (9M)

(b) Which critical aspects of the climate-soil-vegetation balance need to be examined when planning agricultural developments in previously undisturbed areas? (4M)

(c) State the possible ecological consequences of:
(i) bringing forested land into cultivation,
(ii) bringing marginal land into cultivation.
(2 × 2M)

(d) Explain how the pest-predator-parasite relationship may be disturbed by the use of pest controls. (3M)

(e) 'The greater the departure from the natural ecosystem diversity the greater the potential damage from pests and diseases.' Explain why this is often the case. (5M)

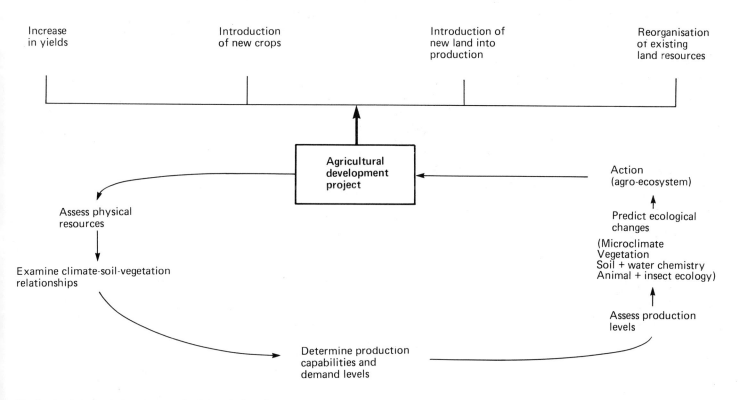

Ecological principles for agricultural development

Inexhaustible	Exhaustible but renewable	Exhaustible but irreplaceable
1.	1.	1.
2.	2.	2.
3.	3.	3.

Table 1   Assessment of physical resources

86

(a) Describe briefly how you would undertake a quadrat survey of vegetation cover along transect line A – B shown in Figure 1. The survey should include *eight* sites. Pay particular attention to the problems involved with the use of quadrats for vegetation sampling. *(6M)*

(b) Explain briefly how you would attempt a more detailed analysis of the vegetation in zone II. *(3M)*

(c) Suggest possible reasons for the stunting of the oak forest in this zone. *(3M)*

(d) Study the three scatter graphs shown in Figure 2. (i) State which graph shows, (*a*) a negative correlation, (*b*) no correlation at all (*c*) a positive correlation. *(3M)*

(ii) Suggest a factor to fit each of the patterns shown on the graphs, stating a reason in each case. *(6M)*

(e) Study the table in Figure 3. Describe a statistical test that a student might use to compare the validity of his/her results (the observed distribution) with the results of previous vegetation surveys on the slope (the expected distribution). *(4M)*

The results of an A-level student's field investigation on a Dartmoor hillslope

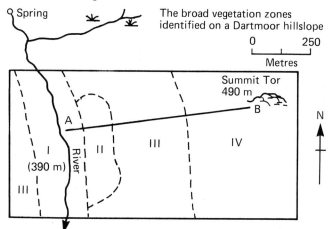

Figure 1

Broad vegetation zones:

I : Peat bog : Sphagnum Moss: Bracken
: Soft Rush Grass: (Juncus effusus)

II : Stunted Oak Forest : (Quercus robur)
Bracken : (Pteridium acquilinium)

III : Gorse : (Ullex galli)
Ling : (Calluna vulgaris)

IV : Open moorland : Ling : (Calluna vulgaris)
Common Bent Grass : (Agrostis tenuis)
Whortleberry : (Vaccinium myrtillus)
Various mosses and lichens

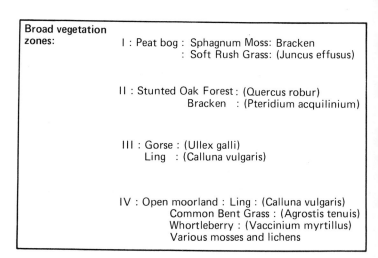

Figure 2

Percentage distribution of Bracken at 8 survey sites along transect A–B

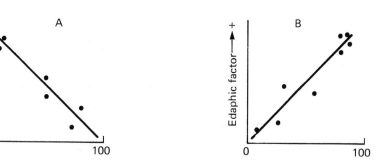

Percentage distribution of Ling at quadrat sites	Quadrat sites							
	1	2	3	4	5	6	7	8
Observed distribution	10	15	15	30	40	45	25	5
Expected distribution	5	10	15	30	45	45	20	5

Figure 3

87

(a) Explain what is meant by the term 'clay-humus complex'. (*4M*)

(b) Explain the interrelationship which exists between the clay-humus complex and vegetation cover. (*6M*)

(c) Account for the loss of nutrients shown at the two points on the diagram. (*2 × 2M*)

(d) Transfers within the nutrient cycle may involve either physical movement or chemical transformation. Explain how these relate to the transfers shown on the diagram. (*5M*)

(e) Nutrient cycles involve either the atmosphere or the lithosphere. Explain why this is the case, stating examples of each of the cycles named. (*6M*)

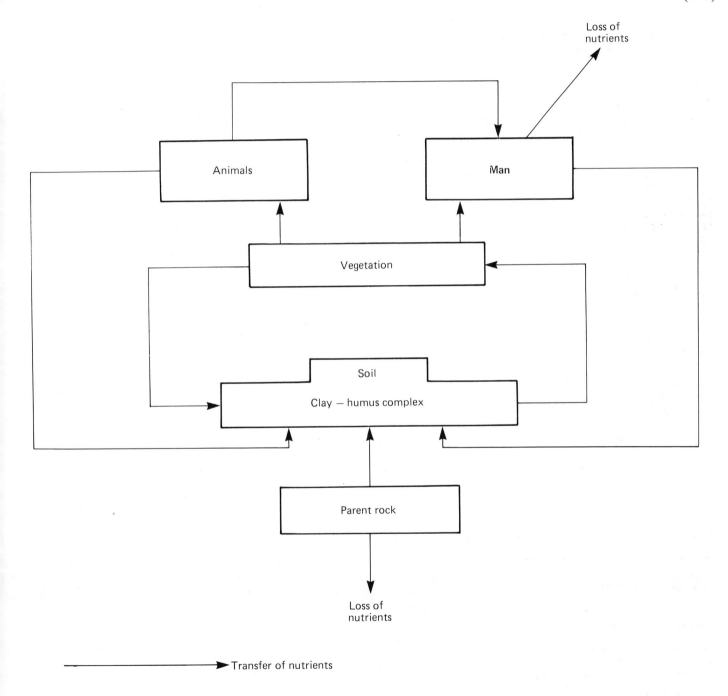

A diagrammatic representation of the nutrient cycle

Acknowledgements

The author and the publishers would like to thank the following for permission to reproduce copyright material in this book: Question 3 – BBC Publications for material from *The Restless Earth* by Nigel Calder, published 1972; Questions 4, 25, 81 and 82 – Aerofilms Ltd; Question 5 – Cambridge University Press for a diagram from *The Study of Landforms* by R. J. Small; Question 8 and 16 – Annals of the Association of American Geographers; Questions 14, 81 and 84 – *The Geographical Magazine*, London; Questions 20 and 71 – Macmillan, London and Basingstoke for material from *The Welsh Marches* by R. Millward and A. Robinson and from *Plants and the Ecosystem* by W. D. Billings; Question 24 – Oxford University Press for a diagram adapted from *Developments in Geographical Method* by B. P. Fitzgerald; Question 26 – Methuen Ltd for a diagram from *Atmosphere, Weather and Climate* by R. G. Barry and R. J. Chorley; Questions 28, 29, 33, 34 and 37 – Thames Water Authority; Questions 39 and 40 – two diagrams reprinted from *Physical Climatology* by W. D. Sellers by permission of The University of Chicago Press, © 1965 by The University of Chicago; Question 52 – Thomas Nelson and Sons Ltd for a diagram from *Modern Meteorology and Climatology* by T. J. Chandler; Question 55 – Hutchinson Publishing Group Ltd for a diagram from *The Climate of London* by T. J. Chandler; Question 59 – Royal Meteorological Society for a diagram from 'The hailstorm' by F. H. Ludlam, from *Weather*, May 1961, Vol. 16, p. 152; Question 65 – Robert M. Basile for a diagram from his book *A Geography of Soils*, published by William C. Brown Co.; Question 69 – W. H. Freeman and Co. for data from *Ecoscience: Population, Resources, Environment* by Paul R. Ehrlich, Anne H. Ehrlich and John P. Holden (© 1977); Questions 74 and 87 – Edward Arnold (Publishers) Ltd for two diagrams from *Vegetation and Soils – a World Picture* by S. R. Eyre; Questions 76 and 83 – University Tutorial Press Ltd for diagrams from *Climate, Soils and Vegetation* by D. C. Money and from *Process and Pattern in Physical Geography* by K. Hilton.

British Library Cataloguing in Publication Data

Guinness, Paul
    Data response questions in advanced level geography.
    Physical geography
    1. Geography—Examinations, questions, etc.
    I. Title   II. Ball, K. H.
    910'.76   G131

    ISBN 0 340 28327 0

First printed 1983
Third impression 1985

Typeset 11/12pt Plantin (Monophoto) by Macmillan India Ltd, Bangalore.

Printed in Great Britain for
Hodder and Stoughton Educational,
a division of Hodder and Stoughton Ltd,
Mill Road, Dunton Green, Sevenoaks, Kent
by St Edmundsbury Press,
Bury St Edmunds, Suffolk.